Creative ideas are like apples sitting on a tree
waiting to be picked.

All You
Gotta
Do Is Ask

**Listen – Cultivate – Till – Fertilize - Pollinate
– Sow & Harvest**

**Why Not Make Your Work Easier and More
Interesting?**

10-16-0 5

By Chuck Yorke and Norman Bodek

ALL YOU GOTTA DO IS ASK

PCS Inc.
PCS Press
809 S.E. 73rd Avenue
Vancouver, WA 98664

bodek@pcspress.com
http://www.pcspress.com

Printed in the United States of America

Printing number
1 2 3 4 5 6 7 8 9 10

The cover is a stereogram designed by Gene Levine

Library of Congress Cataloging-in-publication Data

Yorke, Chuck; Bodek, Norman.
    All You Gotta Do Is Ask

    p. cm.
    Includes index.

ISDN 0-9712436-5-4
1. Production management.   2. Industrial Management.
3. Organizational change.   4. Organizational behavior.
5. Manufacturing processes.

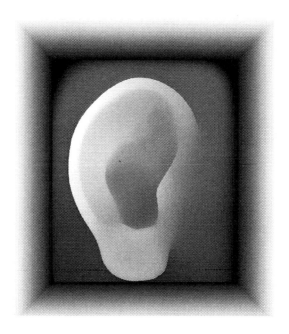

The Cover: The picture on the front cover is a stereogram, a magic eye. Hidden for your discovery is a three dimensional picture (3D) by the artist **Gene Levine**. Relax your eyes, and focus six inches beyond, above or below, the book. You can also look at the cover cross-eyed by holding your index finger a few three inches in front of your nose. Or hold the cover right up to your nose, focus as if looking off into the distance and slowly move the page away. With patience and a little practice you will see the magic happen. For help go to:
http://colorstereo.com/texts_.txt/practice.htm.

ALL YOU GOTTA DO IS ASK

Introduction

All You Gotta Do Is Ask

Why Not Make Your Work Easier and More Interesting?

Workers have knowledge. They're an untapped source of wealth. They have ideas that can improve the company—thousands of ideas that are rarely taken seriously, rarely mined, and vastly underutilized. The organization that doesn't tap into this source is like a poor country village that sits on top of a gold mine.

Years ago American unions had a famous motto, "Work Smarter, Not Harder." Great idea, but it was never practiced. Workers would come to work, do their job as told, and hardly ever were they asked to participate in problem solving activities. Hardly ever were they asked to use their brains.

In this new global market place where we are competing with China, India, the European Union and other countries, we must harness the creative latent talents from every single worker. This book is dedicated to helping you recognize and bring the best from every single employee in your company.

Instead of waiting endlessly for new ideas to emerge, all we have to do is to bring the suggestion box to them, ask for their ideas, listen, and then help implement the powerful suggestions that inevitably surface.

Do this and you will discover the infinite creative potential that exists within every single person in your company. Everyone can and should be involved in solving problems to make their work easier and more interesting. **The highest motivation and learning comes when people use their own creative ideas, and implement those ideas both to develop their skills and to solve problems.**

Taiichi Ohno, former Vice President of Production at Toyota, once saw a worker "sweating." The supervisor was proud of the worker because he was working so hard. Ohno said, "Why not make the work easier so the person doesn't have to sweat? The Toyota style is not to create results by working hard. It is a system that says there is no limit to people's creativity. People don't go to Toyota to 'work,' they go there to 'think.'"

TPS (The Toyota Production System) is now becoming the "Thinking" Production System.

Author's Dedication

"For me, I would say that all you need to define life is imperfect replication. That's it. Life. And what that means is that an entity can make copies of itself but not exact copies. A perfectly replicating system isn't alive because it doesn't evolve. Quartz crystals make exact copies of themselves and have done so from the beginning of the earth. They don't evolve, however, because they're locked into that particular form. But with imperfect replication you get mutants that develop some sort of selective advantage that will allow them to dominate the system. That whole system evolves, and you get this cascade of evolution progressing to more complicated entities. But something preceded all that, something that could do this basic thing of replication and mutation, and that's what everyone is trying to figure out."- Jeffrey Bada, geochemist at the Scripts Institution of Oceanography at U.C. San Diego

The world is forever changing. It is our job to continue to foster that change.

This book is dedicated to bringing new life to all managers, practitioners, students, teachers, and coaches of modern management practices focusing on discovering better ways of enriching the lives of all people at work. This book is dedicated to change – to helping you improve your work environment by empowering all workers to dream, to envision, to solve problems, and to continuously improve and grow on the job.

For over 100 years industry has asked workers to leave their brains at home. Rarely have workers been empowered to participate fully into creative problem-solving activities to make their work easier and more interesting. They have been told to "Just do as you are told and leave the thinking to us."

To be internationally competitive today, a whole new fresh approach to managing has to take place. It is vital that every single worker be challenged to participate by using his or her intelligence and to continuously develop their skills to their maximum capacity. Toyota has done it. In fact, most Japanese companies are doing it and surely you can do it if you apply the simple concepts examined in this book. We imagine as you finish the book and apply these concepts that you will stand back and say, "Wow, why haven't we done this before?"

We want to thank the many people who have helped us to write and publish this book: to Anthony Miriello who dreamed up the title; to Noriko Hosoyamada and Louann Yorke for their insights and support; to our editors: Suresh of the Tao Consulting Group in Chennai, India; Bluefish Bay Editing and Publishing Services; to Alan G. Robinson for his fabulous foreword to the book; Don Dewar, Gary Smuda, Jon Miller, Shelly Blakita, Gayle Shaughnessy, Brian Starkey, and Dale Johnson for their comments; Rhonda Schneider and her team for identifying barriers; Pat Goss and Rob Curtner for their interviews; Mike Karol for

AUTHOR'S DEDICATION

initiating Quick and Easy Kaizen at his company; and Gene Levine for the cover design.

ALL YOU GOTTA TO DO IS ASK

Foreword

By Alan G. Robinson
Professor, Operations Management at the School of
Management at the University of Massachusetts and
Co-author of *Ideas Are Free* and *Corporate Creativity*

All You Gotta Do Is Ask examines how to draw
scores of ideas from each employee, something most
organizations do poorly, if at all. Managers of organizations
are usually either unaware of the power that can be attained
from employee ideas, or they don't know how to tap them.
This easy-to-read book will show you why it is important to
have a good idea system; how to set one up; and what it can
do for you, your employees, and your organization.

I wish more business books were written by a team
like the authors of this one. Norman Bodek has deep,
extensive, firsthand knowledge of best practice in idea
systems around the world. Chuck Yorke is a manager who
played a central role in creating profound transformation in
his organization by applying the principles you are about to
encounter. All too often, business books offer visionary
advice un-tempered by the experience of actually putting it
into practice.

I have known and admired Norman Bodek for almost
twenty years. More than any other single person, he is

responsible for introducing the Western world to lean manufacturing. The simplest way to describe what he did is to say that he was the right person in the right place at the right time. But that leaves much out of a fascinating story.

In the early-1980s, before most people in manufacturing discovered the secrets, Bodek realized that leading Japanese manufacturers weren't just trying harder than their Western counterparts; they were somehow operating *in a completely different way.* He decided to find out for himself what they were doing, and thus became a serious student of the early lean revolution. Over a period of years, he went to Japan often, and sought out the leaders of this revolution. He charmed them. He was persistent in quizzing them and in getting into plants to see firsthand what they were doing. It was he who discovered for the Western world the gurus of the new movement – Shigeo Shingo, Taiichi Ohno, Ryuji Fukuda, Kenjiro Yamada, Bunji Tozawa, and many others. He also identified a number of organizations whose primary mission was to promulgate lean ideas throughout Japanese industry, such as the Institute of Total Productive Maintenance, the Japan Human Relations Association, and the Japan Management Association. At a time when most Western manufacturers remained in denial and were attributing the preeminence of their Japanese competitors largely to external causes (such as friendly government policies, trade barriers, lower wages and the vaunted "Japanese culture"), Bodek saw that their management practices were the real factor in their success. It was immediately clear to him that the radically different Japanese practices would be the future of manufacturing. Companies that followed their example would win out *easily* over those that didn't.

Bodek bought the translation rights to many of the leading Japanese books on lean production and founded

Productivity Press in order to disseminate the new ideas as widely as possible. Today, almost every lean initiative, training course, consultant, and written word on the subject in some way draws on the body of literature that Bodek brought over. He also founded a consulting and training arm, Productivity Inc., to help companies implement lean ideas. Productivity Inc. brought many of the leaders of the lean revolution to the U.S. for speeches, training, and consulting. It ran corporate study missions to Japan that opened the eyes of many Western managers and leaders to the massive changes they needed to make.

One of the distinctively Japanese management practices Bodek found was the *kaizen teian* system – a simple system to gather productivity and quality-improving ideas from front-line employees. In 1989, for example, Japanese companies were averaging more than 37 ideas per employee, of which 87 percent were implemented. Quantifiable bottom-line savings were calculated at more than $4,000 per employee. By contrast, their U.S. competitors put little effort into encouraging employee ideas. In 1989, our *national* average was less than one idea per employee every *eight years,* of which less than a third were actually used. To me, this difference, which is purely a management issue, goes a long way in explaining the distress of U.S. manufacturing over the last twenty years. Over time, the ability to listen to the people who actually *do* the work creates a huge competitive advantage. For example, at the time of this writing, Toyota (which has made employee ideas a top priority for the last thirty years) has a net worth almost as large as the Big Three combined, as well as profits greater than the Big Three combined. The irony is, as this book will show you, that with a little self-discipline and know-how *any* organization can get lots of useful ideas from its employees. It is astonishing how few managers make use of this free and infinite resource.

ALL YOU GOTTA TO DO IS ASK

A book on the subject of *kaizen teian* systems by Norman Bodek alone would be important. But *All You Gotta Do Is Ask* is co-authored by Chuck Yorke, Manager of Organizational Development at Technicolor. I haven't known him for as long as I have known Norman Bodek, but I have tremendous respect for him, too. Chuck attended a seminar of mine in Detroit a year ago, and came up during a break to ask me if I knew Bodek. "He has been working with us at Technicolor, and I think you might be interested to see what we have accomplished." A month or two later, I visited Technicolor and was very impressed. I talked with employees and managers there, and saw a strong system of improvement through employee ideas. As the visit progressed, I knew I was looking at a management team that had realized the fundamental truth that workers can often cut costs and boost revenues *more intelligently* than management can. They are the ones making the products and delivering the services, and dealing with customers and suppliers. They see every day what is wasting time and money, and what can improve the customer experience. From soup to nuts, Chuck Yorke has been involved in the implementation of an effective idea system, and knows the issues you will face. There is nothing more valuable than experience.

Norman Bodek and Chuck Yorke are among the relatively few voices (I count myself in this number too) advocating that business leaders pay serious attention to the ideas of their front-line employees, and learn – as a primary management skill – how to encourage, collect, and process these ideas systematically, continually, and on a large scale.

I hope this book inspires you and leads you to start asking your own people for their ideas. You will become a much more effective manager as a result. Your people will be happier, you (the manager) will be less stressed, and your unit's performance will rise to levels you could not have come close to in any other way.

ALL YOU GOTTA TO DO IS ASK

Author's Foreword

Workers have excellent ideas that they would love to share. Unfortunately, many times their ideas are blocked. This has to change. We can learn much from our workers' great ideas. We must find a way to allow all our workers to fully express their creative thoughts. The only way things will improve; in fact, the only way things have ever improved, is through implemented ideas.

No company is perfect; every organization can be better than it is right now. Just ask the employees. Every day in lunchrooms around the world, people share improvement ideas while they eat their sandwiches. The problem is, when the ideas are brought to management, management has largely been unwilling to listen. Today, with intense international competition, companies need to hear their employees; they need to tap into the creativity of each and every one of their workers.

When workers share their ideas, the conversation usually begins with: "All they have to do is . . ." or "Can't they see that if . . ." or "Why don't they . . ." or "I can't believe that . . ." and so on. Great ideas are discussed, valuable information that can improve work and help the company.

Why has management been unwilling to listen to workers' ideas? Are the ideas a threat, too much of a challenge to managers? Managers might feel that they are the ones who are paid to think and to solve problems, while the worker's job is "just to do the work!" It might be that these managers do not respect the intelligence of the average worker. The system must change. And it will.

We all can learn from each other, because each of us has ideas worth sharing. Management needs to learn, workers need to learn, and business needs to learn. It is important for all of us to be able to express our ideas. The only thing that is constant is the need to change and grow. We always need a better way, and we must find a new mechanism to bring everyone's ideas to fruition.

People have ideas and want to share them. More than that, they want to see their ideas implemented. If ideas are shared at lunch, during breaks, and at the local watering hole, why aren't they shared on the job? How can we capture ideas on the job?

My work experience has allowed me to visit many companies and see how ideas are captured. Typically, it's either a suggestion box on the wall or a 21st century software system with a fancy title like, "The Super Zowie Idea Extraction and Involvement Protocol." In other words, it's a system set up outside of ordinary work to capture ideas. It's a special program and usually we reward people for participation.

Guess what? Programs that are "special" or outside of their normal work are not very successful. We get some results; sometimes we get lucky and capture a great idea. But this is not a normal occurrence.

Rather than something special, we must make it ordinary for people to share ideas. We can create an environment where it's safe, enjoyable, and rewarding to volunteer ideas.

This book will show the way. It will describe how a system can be put into place to capture ideas and the role of management to stimulate creativity within each employee. A process is needed to help people develop and grow, just like a gardener needs to plant seeds and tend the garden. We'll look at cultivating and harvesting ideas as an integral part of business.

We hope you find this book interesting, fun, and stimulating. You can apply our ideas not just with employees, but with customers, suppliers, and everyone else who comes in contact with your business. Be creative, enjoy.

Chuck Yorke

ALL YOU GOTTA TO DO IS ASK

Author's Foreword

As a child I remember playing "tug of war" at summer camp, where two teams of ten campers, holding onto each end of a large rope, would stand on different sides of a small lake. When the bell went off, each team would pull hard to try to pull the other team into the lake. It was a lot of fun. It took a lot of energy. It required real teamwork. And it was vital that each person do his or her share in the pulling. One person either pulling in the wrong direction or not pulling hard enough would cause his or her whole team to lose.

Can we use that analogy with your company? Is it possible for you to be competitive internationally against the Japanese, Chinese, Indians, and Europeans without all employees fully contributing their highest spirit, their maximum talent, and their creative intelligence?

Just read the daily newspapers and you will see more and more jobs going overseas, and more and more manufacturing companies laying off their workers. It doesn't have to be that way if we can somehow get everyone in the company to pull together.

In the late 1800s, Frederick Taylor and William and Lillian Gilbreth created scientific management and the division of labor, which completely changed the way work was performed in our factories. They wanted "flow manufacturing" with people doing very simple repetitive tasks and specifically did not want production workers to "think." "Thinking created stress," stress resulted in tired workers and in lost production. Their method worked at that time. Ford Motor Company, for example, using their concepts became the most successful company in the world and was the forerunner of enormous wealth to American industry.

But nothing lasts forever. While in the past, America was separated by vast oceans from international competition, today with instant communications, jet travel, and speedy ocean liners, foreign companies with "cheaper labor," and "highly educated workers" have taken away over two million American factory jobs in the last few years.

What can you do? Simple. Remove the hidden sign over your factory gate that says, "Please leave your brains here. They will be returned to you after work!"

You can begin to empower every single worker to be involved in continuous improvement by using their brainpower to identify potential problems, solve those problems, set improvement targets and goals, encourage everyone to learn new skills, teach the problem-solving tools, and ask every employee to submit in writing two improvement ideas per month.

In the early 1990s, Woody Morcott, CEO of Dana Corporation, asked all of his employees, 80,000 at the time, to submit two creative ideas per month in writing and wanted 80 percent of them to be implemented. Dana's employees have complied and for over ten years Dana has been receiving close to two million ideas a year to improve the company.

Please have fun reading our book. It is guaranteed

to help you discover simple but very powerful ways to get all employees pulling in the same direction through their own creative ideas. After getting this system started, you will say, "Why haven't we done this before?"

This book will show how a system can be put into place to nurture the creative potential of each employee and the role management plays in establishing and sustaining that system.

Norman Bodek

ALL YOU GOTTA TO DO IS ASK

Table of Contents

ALL YOU GOTTA TO DO IS ASK

Chapter I

A Lesson in Corporate Creativity

*"Every once in a while I feel that I am
at two with the universe." - Woody Allen*

All of us are here for only one reason: to grow. To develop our skills, to participate creatively and make a better place for all of us to live, to be more conscious of the world around us, to be more compassionate and appreciative of life, and to help and serve others in their own attempt to grow.

Each of us has certain lessons to learn while we are here on Earth, and we will be tested over and over to see if we have learned those lessons. The quicker we learn them, the faster we can move on into the vast unlimited horizon. In order for us to grow, we have to overcome our own inertia, our fears of the unknown, our fears of making mistakes and, above all, to overcome our own resistance to change. How can we grow when we want to stay the same and do the same things over and over?

There are many lessons to learn. One of the most

1

difficult lessons is this: how not to react to criticism.

There is an ancient story of a monk sitting on top of a stone wall. While he was sitting on this wall, a distinguished and well-dressed couple walked by and the monk inadvertently dropped a stone onto the ground that splattered mud over the couple's clothes. The man immediately berated the monk for what he did. "You stupid careless person. You have ruined my cloak." The monk looked down, saw what he did and laughed. All he could do was laugh. This infuriated the man even more, who shouted obscenities at the monk. With each obscenity the monk laughed even louder until finally tears came rolling into his eyes and he said, "Please, I do not mean to offend you but I am only laughing because my teacher told me that I needed to be criticized to surrender my ego. He told me, if necessary, I was to pay others to criticize me. I am laughing because you are doing it for *nothing.*

Hopefully, but without offending others, we could learn to be like that monk and look for, accept, and not react when being criticized. For when we do react, we waste energy and our opportunity to grow just stops.

Effective criticism should be carefully expressed. Rather than offending a person's actions, solely focus to help them solve problems. Don't say, "Why did you do that?" This is much better: "Mary, the product has a quality defect; can you think of a way to prevent that from occurring in the future?"

The goal in this book is to offer some very simple, yet very powerful tools and techniques that will help managers, supervisors, workers, and especially you the reader, to open to the infinite creative powers that are available for all of us to grow and to change.

A LESSON IN CORPORATE CREATIVITY

We will describe a very simple process that is working with many companies in the Far East and several American companies. This process will help you open both yourself and your employees to this infinite creative power. You simply ask yourself and the others working with you three questions:

What can we do to make our work easier?
What can we do to make our work more interesting?
What can we do to build our skills, especially our problem solving skills?

Our original journey began a long time ago, at the birth of our universe at the beginning of time, but this story began recently at a coffee shop at the Detroit, Michigan airport. We were passing the time talking about how to get lots of creative improvement ideas.

The conversation included just about everything from the beginning of the universe until now. We spoke about past lives and the Big Bang Theory of how the universe was created. Some of it seemed to make sense, some was speculation, and all of it seemed to flow together as if it should be obvious to everyone on the planet. This story was created by both of us, who will speak as one.

ALL YOU GOTTA TO DO IS ASK

Companies spend huge amounts of money providing training in problem-solving methodologies, leadership development, communication skills, operational tools, team building, lean manufacturing, constraint management, trust, coaching, systems thinking, Interaction Management, root cause analysis, but I wonder how much of it is really put into practice.

A few of these methods and tools are embraced by organizations, by only a few people. The methods are often tried and they work, at least for a short period of time. Many individuals, of course, never try them at all. They sit through a class, accept what is taught, question a little, maybe even laugh and enjoy themselves, then go back to their offices and put the books on a shelf, making sure their personnel records reflect that they'd taken the training class. Then they never use the material. They never once try to see if it could work for themselves or their company. Why?

There's a simple answer: **people are afraid of change.** People are afraid to even try. Many even refuse to admit they are reluctant to change. When people first join a new company they are bustling with ideas and quickly learn to just keep quiet, do their job, and play whatever game is being played.

People realize that they can do what they did yesterday without being criticized. To change is threatening. "If I make a mistake I will be criticized. I will lose my job! To do something new I normally need approval from my boss and when I do go to the boss 90 percent of the time he/she will say no." No is the safest reply, while change gets stymied.

A LESSON IN CORPORATE CREATIVITY

This inability to change and use our new knowledge on the job surely is a cause of frustration for consultants and trainers everywhere.

Why do companies pay for people to be trained and yet are unable to implement whatever that training was for? It is like they are saving the knowledge for some unknowable future. An old saying is that "mañana (tomorrow) is good enough for me, but in reality mañana never comes."

Some people go to the classes just to acquire all of those course notebooks, which then sit on the shelf behind their desks. They think it looks good.

Sometimes, the methods and tools are half-heartedly attempted. Many organizations find it is difficult, if not impossible, to sustain their use.

I was never a great student at school so to compensate for this, later in life, I would buy all of the great books, the classics, and fill my bookshelves with them and read very few of them.

5

ALL YOU GOTTA TO DO IS ASK

Chapter II

But There is a Better Way!

"The best time to plant a tree is 20
year ago. The second best time is now."
- Chinese proverb

Yes, there is a way to help people learn new ideas and then take those ideas and implement them. As you continue reading on you will have the opportunity to discover a fresh approach to empowering people to be more creative and more involved with continuous improvement activities.

Today, leaders often talk about lean manufacturing, Six Sigma, and continuous improvement, yet very few companies have found the easy way to get everyone involved in the process. As an example, lots of people go through great Kaizen Blitzes—they get a surge of energy, make lots of improvements, and then (just like the bunny rabbit that lost to the tortoise) they stop, admire what they have done, and quickly fall back to their old ways. Like putting up a picture on the wall and saying to themselves, "See what I did!" But the big guys, the Toyotas, never stop improving. They are relentless. If you don't wake up soon, they will bury you.

At the airport, people walked by, some with luggage, and others with family or business associates. Some went into the coffee shop, others into the bookstore or newsstand. Some were casually strolling, while others hustled to the gate or to the ticket counter. It seemed that each person had a destination or an objective, such as catching a plane or finding something to eat. While I watched, it dawned on me that while all these people had their own goals, they acted independently of one another. They were all participants in the travel process.

As we talked, notes were jotted down on a napkin. (I am always taking notes with the thought that I might be learning something new.)

Circles and arrows connected the Big Bang theory of creation, with the possibilities of the great things that could be, and how we might deal with the insanity of what was happening in the world right now. We also spoke about the infinite creative potential resting within each of us, and how it was possible to open that creativity and allow it to serve people's personal development and others around them. (Aha, something can be done.)

The conversation started around business issues; and when we spoke about continuous improvement, a transformation began. Instead of the narrow focus about business concerns, talk shifted to the Big Bang theory of how our universe was created. We went from the past eight hours to billions of years ago speeding back to today, and then looking forward at the possibilities for our future.

Hidden Treasure

Countless books, articles, meetings, training programs, and water cooler conversations center around issues like: What is the role of management? How can I effectively manage? Middle management seems like a barrier. How can we empower workers? Etc. etc. etc.

Senior managers sometimes feel their role is to determine what and how things are done, from the strategy of the company to creating and delivering the product or service.

Middle managers feel that their role is to communicate direction from above, to monitor and control the activities of the workers, and to respond to the concerns and problems from below.

9

Workers provide the service or build the product based on the systems, rules, and work instructions created by management. This is how companies operate. It's the best way for companies to function. They are the behavior patterns that work, and make everyone happy. Things are done the best way possible . . . or are they?

Alan Robinson, co-author of the books *Ideas Are Free* and *Corporate Creativity: How Innovation and Improvement Actually Happens* states that approximately 75 percent of the ideas that bring better products and services or bigger savings comes from front-line workers, not from the executive suite. Imagine, even without an organized creativity process, people's ideas do get through. But also imagine if you had a conscious creativity process what amazing new things could happen!

How do companies discover all of these better products and services when front-line workers are rarely asked for their creative ideas?

The goal of this book is to help you recognize the creative talent within yourself and each of your associates and to provide a simple but powerful vehicle to bring out those latent ideas.

The challenge is to discover those great ideas sitting there in your office or out on the factory floor. Managers and supervisors need a new magnifying glass to look anew at what their people are capable of doing.

"...anybody can say anything they want without getting in trouble. Once you've gained that kind of respect from people then you got yourself a great operation." – Pat Pilleri – former plant manager, Dana Corporation

It turns out that senior managers don't always know exactly what's going on. Middle managers often don't understand why the behaviors and thinking that worked for them in the past doesn't work well now. They feel accused of miscommunication, modifying messages from above, and filtering needs from below.

Workers feel accused of not knowing how to do their jobs. They know the problems and challenges to getting the job done, and at lunchtime while sitting with their co-workers, they discuss the solutions to the problems. They say, "Why can't they do it the right way? What a bunch of jerks. I wouldn't tell them; they wouldn't listen to me anyway. I gave them an idea once and the supervisor stole it from me. You wouldn't catch me submitting an idea again. And I am not paid to do their job for them."

If the workers can see the problems and the answers, then why can't management? In truth, workers are the real experts in their 24 square feet of space, but we rarely ask them for ideas. Workers don't really feel empowered, even though management often feels they've been empowering their workers for years.

Toyota's success comes from their "Two Pillars":

1. Their focus on the elimination of all wastes.[1]

2. Respect for the individual – empowering all employees to participate their creative mprovement ideas.

---

[1] Waste is that which does not add value to the product such as defects, carrying inventory, excess motions, moving things, waiting time, etc.

[4] Value added time is the time that customers are willing to pay for, the time we are converting raw materials to finished goods. The opposite is non-value adding time, or wastes, moving things, defects, etc.

ALL YOU GOTTA TO DO IS ASK

Respect for the individual is the first step toward becoming a world-class 21st-century organization. For the past 100 years, people have come to work to do repetitive tasks. They are rarely asked to participate with improvement ideas. The reason they do their jobs is because they've "had to make a living somehow." Ignoring their input might have worked in the past, but it will no longer work in the 21st century. To compete today internationally, we must learn how to harness everyone's creativity.

When a worker detects a problem he may ignore it, hide it, or report it to a supervisor. Supervisors often feel that problem-solving activities are their responsibility and that workers should go back and do their jobs. The supervisor typically will solve the problem even though the worker has the best knowledge about his job, knows exactly what happened, and could easily participate in the problem-solving activity.

Toyota has changed all that! When a worker detects a problem at Toyota, they will pull a cord or press a button and lights and buzzers go off to alert the supervisor and fellow workers that they need help. The line will stop and all workers will stop working. Toyota allows workers to stop the line to get to the root cause of the problem, to solve that problem so that it will not occur again. Toyota recognizes that all workers should participate in problem-solving activities, that problems should be solved exactly when they occur, and the worker doing the job has the best knowledge to use at that time.

Chapter III

Is Respect Missing?

*"Arrogant managers can over evaluate
their current performance and competitive
position, listen poorly, and learn slowly."*
---John P. Kotter in <u>Leading Change</u>

We need a foundation of trust. Managers coach every chance they get. Leaders preach and "encourage" involvement. Companies have recognition programs that try to enhance their worker's self esteem. Managers are trained to communicate more effectively. Organizations send their managers to take classes in time management, communication management, leadership, listening skills, speaking skills, performance planning, valuing diversity, project management, and team building. Some are even sent to "charm school." Still, workers don't feel involved or empowered.

Maybe we just don't ask the workers, "What do you think?"

Many workers spend eight hours a day doing very

simple repetitive tasks in the factory, in the office, at fast food restaurants, at the supermarket, etc. Work is often designed for "monkeys," not humans. We have over-engineered work. While people are often willing to accept mindless jobs in order to make a living, they find their creative fulfillment elsewhere; at home, sports, etc.

Because the work is simple, managers sometimes stand back and think, "Anyone that does that kind of job can't be too bright!" Respect is often lacking. We forget the schooling these people have had.

In the late 1880s, Frederick Taylor wrote about "Scientific Management and the Division of Labor." He separated planning/thinking from working. The manager would do the planning, and the worker would do the work. And then work was simplified into very simple repetitive tasks.

Frank and Lillian Gilbreth looked at worker's duties and attempted to improve the process. They wanted work to **flow** like a river of water. They found that when the worker stopped to think, that tension and stress would take over and at the end of the day the worker was tired and stressed out.

Both Taylor and the Gilbreths took thinking out of the job. Workers were there to do those repetitive tasks over and over again. It was the supervisor's job to think.

I remember my first visit to a cable manufacturer inCon-necticut. I was invited by the plant manager to see if I could suggest ways for them to improve the quality of their products. As I walked around the plant, I noticed a young lady working at a machine that allowed her to spin two cables at the same time. She was fast and highly skilled.

I was introduced to the operator and began talking to her. I questioned her about the quality of the cables being produced. I asked her if she had ever found any defects. She immediately replied that when she first came to the plant around nine years earlier, she noticed some defects on a cable and immediately stopped the machine. She then wrote up a red tag and placed it on the cable and went back to work spinning more cable. She said that a short while later her supervisor came by, noticed the red tag, yanked it off the cable and said, "What are you trying to do, take away the job of the quality manager? Your job is to spin the cable and his job is to find the defects."

So this young operator learned

a hard lesson. It is true that most often employees without Kaizen training suggest things that are out of or beyond their area of authority. In other words, they suggest things for their boss or for others to accomplish, not for themselves. Of course, there is a place for those kinds of activities—to get important things accomplished. However, what we want to emphasize in this book is that we can improve those things within our own area of responsibility, and that everyone has the ability to make changes happen.

The situation this cable operator found herself in existed in many companies as quality once was left in the hands of the quality manager only. But today, quality improvement is everyone's job. Under Kaizen, we want everyone to be responsible for quality improvement.

Toyota changed the way companies solve problems, but it took almost 100 years to get there. Toyota modernized the suggestion system and empowered the worker to solve problems. At one time, Toyota was receiving over 100 small written ideas per worker per year.

Toyota also invented "flow manufacturing" (often called Lean or JIT). They wanted workers to do their jobs and not to think. They didn't want to break the flow. They wanted maximum productivity, except when a worker detected a problem. Then they wanted the worker to stop working and fix the problem immediately. In fact, they allowed the worker to stop all of the other workers in the factory from working.

Wow, what incredible respect to give workers; the power to stop the entire plant. I appreciate that. They wanted the worker, the one with the best knowledge of the problem, to solve it immediately.

Allan Kempert, Quality Assurance Supervisor, was part of a problem-solving team investigating parts not matching the blueprint. While investigating the issue and speaking with the operators, he soon realized that the operators were "just doing what they were told." The operators were told the dimensions to check and where to record them. They were not shown how to match their part to the drawing. During the problem investigation, the department manager said, "How much do we want them to know? Should we have them match against the blueprint? People shouldn't have to be engineers to do this job."

This helped Allan realize that daily he needs to listen to what the operators are saying. For years they had been telling him, they were, "just doing what I was told." He didn't realize until this problem developed that it meant the operators were only told the dimensions to check and where to record them. They were not told how to match items to the blueprint. To solve this problem, my friend trained the operators on specific things they needed to know about matching parts to prints.

Henry Ford applied Taylor's principles in his factories. Engineers did time studies to find the most efficient way to do work and managers told the workers how to do the work. Ford said that if something doesn't add value, it's a waste. Waste is anything the customer isn't willing to pay for.

Taichi Ohno, former Vice President, Production at Toyota, visited Ford's Rouge complex. Ohno believed that if something doesn't add value, it is waste. During his visit, Ohno noticed something more. Workers were passing defects along the line. This created rework and the necessity of keeping extra inventory along the line. This was waste, "muda," as Ohno called it, and muda was unacceptable.

Ohno determined that the worker who does the work must identify and eliminate waste on the production line. Where Taylor felt workers were "doers" not "thinkers," Ohno felt the worker needed the ability to control the work process, modifying the work even if it meant stopping the line. Workers needed to be "doers" *and* "thinkers."

In contrast to Taylor, Ohno felt that thinking was too important a task to be left only to managers.

So in Ohno's world, the worker not only needed to know how to do the work, he needed to know how to identify problems, think of solutions, and have the ability and the responsibility to modify how the work gets done. The manager must trust the worker's judgment. This can be very threatening for a manager who feels he needs to tell the worker what to do and how to do it.

Dr. Shigeo Shingo, co-creator of The Toyota Production System, also known as Lean Manufacturing, went one step further. He invented the Poka-yoke system.

IS RESPECT MISSING?

He asked workers to build very simple devices to absolutely prevent defects from occurring. He gave a one-day course to a manufacturing plant and one year later, the employees had produced hundreds of Poka-yoke devices.

Haruo Shimada from Keio University in Japan calls the Toyota Production System "human ware." Some people are now calling the Toyota Production System the Thinking Production System. It is a system that maximizes productivity, as well as personal development and motivation through a combination of worker participation, management style, and technology.

With "human ware," people rotate jobs every couple of hours. Compare that rotation with companies where workers do the same thing over and over again for the entire day. A worker once said about repetitive work, "I would go bananas if I had to sit at a press all day and spit one part out after the other." The response to him was, "Some people like to do that. It gives them an opportunity to think about what they are going to cook for supper."

Fujio Cho, President of Toyota, tells employees, "Always think about what your next step will be. Continuously improving yourself is the mark of a true professional."

Recently I discovered a video of Dr. Shigeo Shingo, co-creator of the Toyota Production System, while he was consulting on the factory floor of an American company. Dr. Shingo watched and listened to manager's present problems created by inattentive workers. The conversation went on for around ten minutes. The managers never considered asking the workers what they thought about the errors that were occurring. Workers were sitting in front of the camera being totally ignored, as if they weren't even present. Managers looked at the workers as if they were

19

ALL YOU GOTTA TO DO IS ASK

just extensions of machines and never used their brains.

Chapter IV

What Is Wrong?

*"The medicine I am prescribing for
you is a miracle drug and very powerful, but
there is one problem with it, the medicine won't
work unless you take it. I may tell you
wonderful things, but you're not going to be
successful unless you actually do what I am
telling you. There are a lot of people in this
world who worry that the medicine might be
bitter or might have side effects, and who find
excuses to avoid swallowing it. Behavior like
that will never lead to success. Are you ready to
take your medicine?" – Shigeo Shingo*

What is wrong? In addition to frustrated managers
and unhappy workers, we live in a world where the
customer wants things better, faster, and cheaper. When
ways are found to cut cost, reduce delivery time, and
increase quality, the customer still wants it better, faster,
and cheaper. Thus, we search for answers to accommodate
our customers.

In an attempt to find answers, we measure things.

Expenses are converted from fixed to variable. Engineers look for more capacity, maybe better computer technology. We use project management tools and chart out the time it takes to accomplish tasks. Work becomes standardized. Process capabilities, financial performance, delivery performance are measured. Whether we do this internally or we outsource it, our infrastructure grows.

Sometimes senior management becomes frustrated, shuts the door, and sends the work off to China. All this really does is postpone the problem. While work is sent to China, Toyota and other Japanese companies come here to build new plants to work with the same American workers, and still compete very effectively with those products made overseas. Japanese managers see something in American workers that we refuse to see.

Lean Thinking

Let's reflect again on Toyota. In 1960, Toyota, like other Japanese companies, had a reputation for making "junk." How did Toyota go from making junk to producing the world's highest rated quality automobile—the Lexus? It was not by developing one, or even 1000, new technologies. It came from two things: The Toyota Production system now called Lean Manufacturing and from the continuous improvement ideas of every single worker.

One powerful aspect of the Toyota Production System is giving all workers the right to stop the line when they detect an error, a defect, or a potential problem. At the Toyota plant in Georgetown, Kentucky, the line can be stopped hundreds of times in each shift. As soon as the line stops, colored lights go on and buzzers ring alerting other employees and supervisors that someone needs help. It normally only takes a few seconds to detect the cause of the problem and to solve it.

It is amazing to give each employee the power to stop others from working. It proves how serious Toyota is in producing defect-free automobiles and empowering workers.

Remember the story of the Tortoise and the Hare? America is like the rabbit. At the end of World War II, the United States was light years ahead of everyone else. America, like the rabbit in the story, rested at the side of the road, occasionally waking up with some new technology to jump ahead. The Japanese, like the turtle, plodded ahead each day, inching forward with the ideas of all of their workers.

Toyota had two great geniuses, their principal guides that moved them along. Dr. Shigeo Shingo, Toyota's brilliant independent consultant (their Sensei, or teacher) and Taiichi Ohno, Vice-President of Production, both of whom created and developed the Toyota Production System (Lean manufacturing).

The heart of this system was bringing the suggestion box to the worker - getting ideas from everyone on a regular basis.

Shigeo Shingo structured and popularized the Toyota Production System. Taichi Ohno co-conceived the system and enforced it at Toyota. James Womack, author of The Machine That Changed the World, gave it a new name, Lean.

The Toyota Production System is a great jewel developed and perfected by Toyota, but it is only one of many. Toyota's product development system is another jewel. It has been described in a couple of articles in the MIT Sloan Management Review, yet it has not been as widely embraced as Lean manufacturing.

Toyota's people involvement system is another incredible jewel that until recently has not been widely embraced outside of Japan. People are involved at Toyota through their quality circles program and their suggestion system. Of course, many companies have quality circles and suggestion programs, but with a different focus.

Quality circles and suggestion programs are used to solve problems and improve the process. Quality circles are groups of workers who meet regularly to discuss problems in the workplace and find solutions. Employee participation is used to increase the quality of work. Dr. Kaoru Ishikawa, who created quality circle activities in Japan, believed the purpose of circles was to train people. Toyota uses its quality circles and suggestion system to develop and empower its employees to grow and take on greater responsibilities.

What should America do? Continue investing in the latest management fad or another consultant; or should we become virtual companies and outsource the manufacturing overseas to China or India? Something needs to be done fast as budgets are squeezed and infrastructures balloon, and improvements are no closer than before. Pain surfaces throughout the company, and our improvements fall short of expectations with little or no impact on the bottom line.

Companies such as Toyota offer clues. The good news is that their practices actually work.

Chapter V

The Power of the Suggestion Box

"It is well to remember that the entire universe, with one trifling exception, is composed of others."- John Andrew Holmes

One method used in the past to encourage involvement was the suggestion box. Some companies still have them hanging on the wall, collecting complaints, dust, and maybe a rare good idea. At some companies, suggestion boxes are merely wall decorations. Other companies have taken them down since they received little activity and most of the suggestions weren't acted upon. Those that do have active suggestion systems often find that the ideas submitted are for someone else to implement, not the person submitting the idea. This puts a large burden on management to get the idea executed. If you listen carefully you might hear, "I have enough work to do; I don't have time to implement your idea!" And frankly, people do not like to carry out someone else's idea. In America according to the Employee Involvement Association, the average worker submits in writing one suggestion every seven to eight years. Toyota, at one team, received over 100 ideas per year, per worker.

It All Began at Kodak in 1898

The British claim that the suggestion system was born at a Scottish shipbuilder, William Denny and Brothers in 1880, but I like to credit Kodak who in 1898 started the first American suggestion system. The first idea was "Clean the window!" That was a very good idea. Lighting was minimal at the time. The worker rose to the occasion. The supervisor may have thought the idea was just an extra burden for him. So a great employee empowerment system, an involvement system, was quickly squashed. Instead, the normal cost saving suggestion system emerged whereby the worker is given 10 percent of the savings. The problem is, as soon as you give away any of the company's money, accountants become a part of the process to kill it. It is their task to precisely calculate the savings. This usually takes so long that it in turn kills the system. You can't blame the accountant because that is their job. (Imagine the company keeps 90 percent, but accountants can spend months making sure that 10 percent is correct - simply ludicrous.)

I read recently where a person working for the U.S. Navy was told that his idea was accepted. The problem was that it took seven years for him to get any feedback.

We know that we must do things better, faster, and cheaper. We need to continuously improve. A program may be put in place by hanging a box on the wall and asking the workers to be involved. It seems like a good idea, but it doesn't work. We have the same problems, no solutions, an unhappy workforce, and frustrated managers.

It was December 2002 when United Airlines asked all of its employees to sacrifice and take a reduction in pay to avoid bankruptcy. All their employee's unions, except the mechanics, agreed. The mechanics did not like the cut in vacations and were unhappy with the quality of work life. What could United do to improve the quality of work life? Management would provide clean bathrooms, better tools, etc., but they could not control the weather, nor change the airplanes very easily. What exactly did the union mean by quality of work life and what could United do to please them? Very simply, United could ask the mechanics to come up with improvement ideas to make their work easier and more interesting and then listen to their ideas and allow the mechanics to implement them.

Instead of imposing things from the top to try to make things better, simply ask and empower the workers to become more responsible for their own well-being. Of course, management can watch, advise, and govern how the ideas of the workers are implemented. We are not giving the keys to the workers entirely. Rather, they are being asked to participate in making the environment better.

A manager's perspective is limited. United, for instance, has thousands of workers all over the world and senior management can see and control only a fraction of the problems. When the entire work force is enlisted into an improvement process, thousands of eyes are looking improvements, and it is simple to control the entire process.

As the old saying goes, "If the mountain won't come to Mohamed, then Mohamed must go to the mountain." The way companies are managed has not kept pace with the speed at which the world has changed. If employee suggestions are a good idea and employees won't come to the suggestion box**, then the suggestion box must be brought to the employees.**

ALL YOU GOTTA TO DO IS ASK

To survive today in this fiercely competitive world everyone's ideas are needed.

It is a terrible waste companies no longer can afford - the waste of not using people's brains at work.

In fact, the only way to get continuous improvement on a day-to-day basis is to enlist all workers in improvement activities. It can be done. Just ask a Dana Corporation manager who receives, on the average, two ideas in writing per month, per employee. I'll repeat that. They get *two ideas in writing per month per employee.* It almost sounds unbelievable. When you see how simple it is to administer, you will kick yourself that it wasn't done before.

Two Ideas Per Month

When I told a manager in another company that Dana gets two ideas per employee per month, he just couldn't fathom it. His mind stopped listening, for he had 350 people in his plant and knew that he could not manage 700 ideas per month. Then I explained to him that what makes the system work is that the manager does not have to manage those ideas. The manager empowers the workers and their immediate supervisors to manage the ideas. The manager receives daily improvement reports and sits back and watches with glee. His company improves with happier employees, greater quality of life, cost savings, quality improvements, safety improvements, and customers that are more satisfied. It sounds like utopia, and it is. The dollar savings are beyond what most people would believe.

You should be saying to yourself, "If I don't jump in and embrace this technique, my company will never be able to keep up with competitors. When I receive only one idea every seven years from a worker and Dana is getting two per month, the divide between us will become enormous." This is how Toyota transformed from making "junk" in 1960 to becoming the world's leader in quality. They simply asked every worker to solve problems and look for improvement opportunities in their own work area.

The key is that this is a continuous improvement system where immediate supervisors, middle managers, and others come directly to each employee and learn directly from the employee about new ideas. People don't voice constructive and meaningful ideas to a box on the wall, but they will to a supervisor or manager who proactively solicits ideas from them.

The secret, of course, is for the supervisor to listen to the ideas from their employees! Those same problems and challenges and lunchtime solutions shared with co-workers can now be ideas that are acted upon.

A few of the benefits include:

- Building a foundation of trust;
- Encouraging involvement;
- Enhancing our worker's self esteem;
- Communicating more effectively between managers and workers.

There is much more payback than what simply appears on the surface by taking the suggestion box to the employee.

ALL YOU GOTTA TO DO IS ASK

Chapter VI

The Box on the Wall

"The wireless music box has no imaginable commercial value. Who would pay for a message sent to nobody in particular?" - Response to David Sarnoff seeking investment in radio

When a suggestion goes into the box on the wall, what happens? Usually it is given to the manager of the department that could implement it, or worse, to a suggestion program committee. Usually the person who submitted the suggestion receives either no response, slight feedback weeks or months later, or reasons why the suggestion can't implemented:

- "We don't do it that way";
- "There's no budget to do this";
- "We tried it before and it didn't work";
- "It won't work";
- "We don't have the time";
- "We will never get approval to do that."

There are many excuses, but they all mean, "We

don't want to take the time" or "We don't care," no matter how the response is worded. The burden is passed to others, and everyone has too much to do already.

How do things change by taking the suggestion box to the employee? We reinvent the continuous improvement process from the bottom up. We create a straightforward approach to redesigning our processes, reducing costs, and increasing efficiencies. It sounds pretty ambitious. Let's see how it's done.

Managers start by sincerely recognizing that each person has the potential to contribute ideas in solving problems that relate directly to his or her work. Supervisors, middle managers, and others can go to employees individually and learn directly from them about their ideas.

Look at each employee as if he or she is the expert on the job and you want to learn how to improve the company by listening to the expert. Ask the workers very simple questions, such as:

- "What ideas do you have to make your work easier?"
- "How can you make your work more interesting?"
- "What can you do to build your skills?"
- "How can you make your customers happier?" (The
- customer being the next person that receives your
- work);
- "How can we cut costs?"
- "How can we make your job safer?"
- "How can we improve the quality of our products?"
- "How can we improve the process?"

And you listen.

Mistakes

Whenever a problem occurs or a mistake is made, **don't criticize.**

Don't call them "stupid." Don't say, "Why did you do that?" Once the mistake is made, you can't go back in time. You can't undo it. Learn from that mistake. **In fact, the best way to learn is from our mistakes.** Treat a mistake as an opportunity to learn. Of course, we also hope that the mistake will not occur again. When a mistake happens, stop; ask the workers to review, like scientists, what caused the problem and ask them to find a solution in the future to prevent the mistake from happening again.

Mistakes are jewels! Unfortunately, our school systems had it all wrong. When a child made a mistake, they reduced that child's grade. Children were punished for their mistakes. The natural reaction to this punishment is to hide mistakes. "I didn't do it."

Imagine a totally new system that when a child makes a mistake the teacher will stop, recognize it as a very powerful learning opportunity, ask the child who made the mistake to discuss it and ask fellow classmates for their ideas on how to prevent the mistake from reoccurring – to look at it as a learning opportunity and not a moment to embarrass the child.

If management continues to punish for mistakes made, employees will not come forward with ideas for improvement. If innovative ideas from the workforce are wanted, management must embrace reward. They must take the suggestion box to the employee.

Once again referring to Toyota when a mistake happens, a problem or potential problem is detected, the worker has the power to stop the line, correct the problem, and try to get to the root cause so that the problem will not occur again.

Generally, ideas people share fall into two categories: ideas for solving their own problems or improving their processes, and suggestions for improvements for others.

What does the supervisor do with these ideas?

When suggestions are for *others* to do something, the supervisor must ask the employee to write up the suggestion, or assist in writing it up if necessary, and take it to the manager/supervisor of the area affected. Often in organizations, when we start this new system of bringing the suggestion box to the worker, people come up with ideas for others to do something, not themselves. Don't reject those ideas. However, it is far better when people come up with small ideas that they can do themselves or with their work teams. It might take a little time to get started, but people will respond and learn the rules of this new game. You can help them look around their work area and ask them simple things:

- "Does the pipe on the floor pose any potential problem?"
- "You bend often. Is there a better way to get the material you need?"
- "Can you organize your tools so that you can find them sooner?"
- "How can we improve the value added time?"[4]

THE BOX ON THE WALL

When the employee suggests an improvement in his or her own area, if necessary, the supervisor or the worker's team members should help the worker implement the idea. The key is to have employees implement their own idea or to take the leadership in seeing that the idea is applied.

It is helpful to videotape the process where the mistake was made and ask a group of engineers and workers to view the tape and come up with new ideas to improve the process. The book, _The Idea Generator— Quick and Easy Kaizen,_ by Bunji Tozawa and Norman Bodek, is a manual for implementing this process. The process enrolls the employee in the improvement efforts and shares the improvement with others in the organization.

Working with one company, after two of years we went from 13 written ideas a month to 2,800 written ideas a month. Imagine how people felt when their ideas were implemented, and it worked, and they received recognition from their bosses and their fellow workers! And the company saved over $10 million from those ideas. Wow!

We've taken that dust-covered suggestion box off the wall and given it some legs. Now we can capture and implement improvement ideas. But what about all those other issues?

Managers and supervisors are now going to each employee individually and learning directly from employees about their ideas. They are now asking and listening carefully. This is valuable for a few reasons, not the least of which is that problems can best be solved where they are encountered, not in an office removed from the workplace. The employee and the supervisor are communicating, working together, and can work on implementing the employee's idea where the employee works. Who knows the work the best? In most cases, it is the worker, who does the task every day. In the past, we rarely ever asked the employee how to improve his or her work.

While the universe was growing and expanding each moment, something new was happening. We once lived in caves. We didn't always have television sets, mobile phones, and fast automobiles. It might have come slowly, but creative ideas are fundamental to the expansion of the universe. Take your mind back to the year 1904, look around, and then look around today. Where did all of the advancements come from? Of course, they came from our minds, our openness to be a creative vehicle for the expansion of this energy, from our ability to solve problems, and from our ability to try new things.

Now think about all of the problems around you today: too many people, air and water pollution, diseases, too many languages, not enough food, the homeless, etc. Just think about the problems of today. How will they be solved? The same way—this unending process using our creative minds will solve them.

At work today all levels of management must be more involved. What better way is there for any manager to find out what is really going on than to speak with the people who do the work, where they work; and listen to problems, solutions, and improvement ideas. Employee's perception of management is based mainly on the behaviors exhibited by their managers and supervisors. What better behavior can a manager exhibit than by listening to people where they work, encouraging their input, and involving them in implementing their ideas.

Frederick Taylor and the Gilbreths in the late 1800s and early 1900s started scientific management by dividing tasks, simplifying work, and making people much more efficient.

Prior to Taylor, a worker did the complete job of making a product for the customer. To build a chair, for instance, a carpenter would meet with the customer and help pick out the size, shape, color, and kind of wood. Then the carpenter would cut the wood, sand it, paste it, polish it, and assemble it exactly as the customer wanted it. There was a totality to work. There was variety. There were creative opportunities.

Taylor separated the planning from the actual work. Management then did the customer contact and the planning, and work was enormously simplified for the worker who did the work.

The Gilbreths noticed that when people hesitated and stopped to think, stress took over and they became more fatigued. They wanted work to flow, and thinking broke the flow. After applying the Gilbreths' methods, a bricklayer was able go from 600 to 1700 bricks per day. But the result was a division of labor; management did the thinking and workers were told to "leave their brains at home." At the time this method was powerful for industry. Productivity vastly improved and through simplifying work, machines were then able to automate people's skills.

Using the work of Taylor and the Gilbreths, Henry Ford was able to build an automobile in just four days, from iron ore coming out of the ground to the finished car being placed onto the railroad train outside of the factory. And he was able to double the worker's pay. The down side, though, was that his workers ended up doing repetitive tasks.

The transition from highly skilled people to repetitive automation brought great wealth to the industrialized world, but it also brought with it boring, unfulfilling lives at work. On the one hand, society benefited from inexpensive goods, while on the other hand, people suffered by doing boring, unfulfilled, uncreative work.

It has taken 100 years to realize that we can have both highly automated factories and places where people can be productive and creative.

Since hesitation brings fatigue, we want a worker to continue to work in flow without thinking. We want him or her to be highly productive. That is, **UNTIL** there is a problem. Then, we want the person to stop, stand back, and be the prime one to solve that problem.

At times, we do not want people to think because it causes fatigue and mistakes happen. However, we do want people to use their brains at work to solve problems. Unfortunately, it has taken us over 100 years to understand how the two activities can work together.

The old idea that a worker comes to work and leaves his/her brain at home is over. People have brains and they have the ability to improve their own work. All you have to do is start to ask them.

To survive in this highly competitive world we need the ideas from every single person in the company. Jobs are leaving for China and elsewhere for "cheap" labor. To keep up our standard of living we must:

1.   Use people's brains;

2.   Challenge everyone to learn and grow at the workplace.

The best way for a person to grow is from his or her own ideas. Sure, when the boss tells us to do something, it must be done. Most of us really have no choice. Workers are paid to follow instructions. However, when the ideas are ours, we are much more excited and motivated.

When managers do listen to their employees, incredible things can happen! Isuzu Motors, Panasonic, Sanyo, Yamaha, and other Japanese companies claim that they save $3,000 per year, per employee from their employee's improvement ideas. There is an untapped bank out there in your factory and in your office. People have wealth to share with your company, but they are rarely ever asked for it.

At ArvinMeritor Inc. a Troy, Michigan based automotive parts supplier discovered many benefits at one of their facilities:

"Lean buying-in also was bolstered by good

relations with rank-and-file workers at the plant, as evidenced by the facility's subsequent ability to post impressive gains in employee involvement. In addition to the 40 hours of annual training per employee since 1997, the unit has averaged 21 kaizen ideas submitted per employee since 1997. Annual savings per employee is $4,285, at a cost of $204 per employee (corrected) since 1997. The idea implementation rate is 95%, and safety incidents declined 86% since 1999. Incident rate has fallen 48% since 2002, and the lost-time rate/hour has dropped 60% since 2002. The plant's successes have earned more than the Shingo award, as it has been recognized as a Ford Q1 supplier, a Ford Full-Service Supplier, and a four-time recipient of the State of Indiana Quality Improvement Award"

- http://www.sme.org/cgi-bin/get-item.pl?ME04ART48&2&SME&

It takes only a few moments of your time to communicate and encourage employees to be involved. Employee involvement in this way builds trust. More trust encourages more communication and better involvement. These behaviors build on each other. Our behaviors affect our attitude, which then affects our perspective, which affects our attitude, which affects our behaviors. The moving suggestion box can be used not just to generate ideas, but also to positively affect workforce attitude and behavior.

Gary Smuda, a plant manager, remarked after he started to bring the suggestion box to his employees, "Most of us come from past corporate cultures in which managers were the only firemen. Now I have 450 firefighters, and they aren't coming to my door saying, 'We have problem.' Instead they're knocking on my door and saying, 'This is how we fixed this problem,' which is awesome."

That means 450 people are finding ways to serve the customer better, faster, and cheaper through a simple idea implementation system, not accountants converting expenses from fixed to variable, not engineers looking for more capacity, not buying more computers and project management tools. No growing infrastructure, only workers solving their own problems and being involved in improving the organization.

A question managers frequently ask is, "How in the world can a manager handle two ideas per month per employee?" Good question! You don't handle them. Instead, the workers who come up with the original ideas and implement 90 percent of those ideas. Your job as a supervisor/manager is just to say, "Great! Go and do it."

What kind of ideas do these employees come up with? Some are cost saving; others involve safety, quality, customer service, throughput, and morale. Added together, they have a positive impact on the bottom line. Some examples are:

1   Install simple motion detectors in the hotel hallways to turn on lights when someone enters.

⊗ ⊗ ⊗ ⊗ ⊗ ⊗ ⊗ ⊗ ⊗ ⊗ ⊗ ⊗ ⊗ ⊗ ⊗ ⊗ ⊗

2   Damaged boxes are all placed into a large container, which is very heavy to carry. Use smaller containers.

⊗ ⊗ ⊗ ⊗ ⊗ ⊗ ⊗ ⊗ ⊗ ⊗ ⊗ ⊗ ⊗ ⊗ ⊗ ⊗ ⊗

3   When the fan is blowing, covers blow onto

the floor. Tape a piece of cardboard to the side of the machine to cut down the wind.

⊗ ⊗ ⊗ ⊗ ⊗ ⊗ ⊗ ⊗ ⊗ ⊗ ⊗ ⊗ ⊗ ⊗ ⊗ ⊗ ⊗ ⊗

4   A small door does not stay tight enough. Take a piece of cardboard, fold it, and insert into the side of the door to keep it tight.

⊗ ⊗ ⊗ ⊗ ⊗ ⊗ ⊗ ⊗ ⊗ ⊗ ⊗ ⊗ ⊗ ⊗ ⊗ ⊗ ⊗ ⊗

5   Many workers do not know how to run all of the machines. Post a chart showing each person's machine skills, then when we are slow, set up training classes until everyone knows every machine.

⊗ ⊗ ⊗ ⊗ ⊗ ⊗ ⊗ ⊗ ⊗ ⊗ ⊗ ⊗ ⊗ ⊗ ⊗ ⊗ ⊗ ⊗

6   Work area gets dirty. Allow people five minutes before close of day to clean their work areas.

⊗ ⊗ ⊗ ⊗ ⊗ ⊗ ⊗ ⊗ ⊗ ⊗ ⊗ ⊗ ⊗ ⊗ ⊗ ⊗ ⊗ ⊗

7   Line leader sets up the work, but when the line leader is absent, people stand around and wait. Train at least one back up for each line leader.

⊗ ⊗ ⊗ ⊗ ⊗ ⊗ ⊗ ⊗ ⊗ ⊗ ⊗ ⊗ ⊗ ⊗ ⊗ ⊗ ⊗ ⊗

8   When entering the warehouse the doors only swing open one way and you can't tell if someone is standing behind the door. Cut windows into the door to see if someone is standing behind the door.

⊗ ⊗ ⊗ ⊗ ⊗ ⊗ ⊗ ⊗ ⊗ ⊗ ⊗ ⊗ ⊗ ⊗ ⊗ ⊗ ⊗ ⊗

9 Cases are falling off the line. Adjust the rails to make cases go straight down the line.

⊗ ⊗ ⊗ ⊗ ⊗ ⊗ ⊗ ⊗ ⊗ ⊗ ⊗ ⊗ ⊗ ⊗ ⊗ ⊗ ⊗ ⊗

10 Standing up for eight hours hurts people's legs. Place cardboard under the mats to create a softer cushion.

⊗ ⊗ ⊗ ⊗ ⊗ ⊗ ⊗ ⊗ ⊗ ⊗ ⊗ ⊗ ⊗ ⊗ ⊗ ⊗ ⊗ ⊗

11 Put all product weights into the computer so shipping personnel don't have to look up weights each time an item is shipped.

⊗ ⊗ ⊗ ⊗ ⊗ ⊗ ⊗ ⊗ ⊗ ⊗ ⊗ ⊗ ⊗ ⊗ ⊗ ⊗ ⊗ ⊗

12 Use an Excel spreadsheet rather than carbonless copy (NCR) paper for travel vouchers.

⊗ ⊗ ⊗ ⊗ ⊗ ⊗ ⊗ ⊗ ⊗ ⊗ ⊗ ⊗ ⊗ ⊗ ⊗ ⊗ ⊗ ⊗

13 Create files on shared computer drives instead of creating multiple copies and filing in various areas.

⊗ ⊗ ⊗ ⊗ ⊗ ⊗ ⊗ ⊗ ⊗ ⊗ ⊗ ⊗ ⊗ ⊗ ⊗ ⊗ ⊗ ⊗

14 Company box truck does not start on cold days due to plugs not heated up and running down batteries. Run an extension cord from the truck engine warmer to the shop to keep plugs warm.

⊗ ⊗ ⊗ ⊗ ⊗ ⊗ ⊗ ⊗ ⊗ ⊗ ⊗ ⊗ ⊗ ⊗ ⊗ ⊗ ⊗ ⊗

15 Take previously-printed paper and turn it around to use again on the printer. Cut paper

cost by 50 percent.

⊗ ⊗ ⊗ ⊗ ⊗ ⊗ ⊗ ⊗ ⊗ ⊗ ⊗ ⊗ ⊗ ⊗ ⊗ ⊗ ⊗ ⊗ ⊗

16 Knob on tape machine broke off, making it difficult to turn handle. Stick a pen through the hole to replace the knob.

⊗ ⊗ ⊗ ⊗ ⊗ ⊗ ⊗ ⊗ ⊗ ⊗ ⊗ ⊗ ⊗ ⊗ ⊗ ⊗ ⊗ ⊗ ⊗

17 People are standing in the same spot getting sleepy doing the same job all day. I moved everybody every two hours to give them a chance to learn different things.

⊗ ⊗ ⊗ ⊗ ⊗ ⊗ ⊗ ⊗ ⊗ ⊗ ⊗ ⊗ ⊗ ⊗ ⊗ ⊗ ⊗ ⊗ ⊗

Each of the above ideas is small, but all slightly improve the job. Imagine when each person is implementing at least two ideas per month.

Chapter VII

Create the Environment

"The pressure necessary for success is applied by good people working within good systems that empower them to use their skills."—Gerald A. Michaelson in <u>Sun Tzu: The Art of War for Managers</u>

When a worker doesn't behave as we would like, or when a worker doesn't appear to have the proper attitude required for the job, we seek ways to change that person. Unfortunately, many times this occurs in the form of discipline, or threats of discipline. The fact is that the worker most likely is already motivated, just not motivated to do the work we desire. Instead of telling the person what to do to be better motivated, we should ask the person simply, **"What can you do to make your work easier and more interesting?** What can you do to improve the process? What do you see that stands in the way of making your job easier? I will help you improve your job."

On June 24, 1980, NBC aired a television program entitled, *If Japan Can, Why Can't We?* that introduced

ALL YOU GOTTA TO DO IS ASK

Americans to Dr. W. Edwards Deming. General Douglas Macarthur invited Dr. Deming to Japan, to help the Japanese rebuild their industry after World War II. Dr. Deming was world renowned for his role in quality improvement, and the Japanese named their industrial quality award, the Deming Prize, after him.

Dr. Deming believed it is the system of work that determines how work is performed. He stressed that only management can create the system, provide training, allocate resources, and provide the tools and environment necessary to produce superior quality. Deming stressed that the way work is performed is determined by the system. **The worker does have a role, and that is to resolve problems caused by actions under the worker's control.**

The Japanese listened to Dr. Deming and other great quality gurus and made enormous strides to improve product quality. It's a lesson that we in the West have not yet learned.

That lesson is this: *All You Gotta Do Is Ask.* Once you do, then sit back and watch magic take place at the work site. The worker stabilizes the process and produces acceptable output. If the process is not being stabilized, then management must change the system.

Look closely at the behavior of the workers. Addressing repeated behavior issues at an event or pattern level will not provide continuous long-term results. "Why did you do that?" is a phrase often used, but it is confusing to the average worker. If the worker knew why, he or she wouldn't have done it. Only by addressing this behavior at the systemic structure level can a system be developed to provide the desired behaviors.

In the airport, as we sipped our drinks, we spoke about the origination of our world in the Big Bang. Imagine at the beginning of our universe there was no space at all, absolutely nothing existed. The Big Bang theory says that out of a single point, a singularity, energy flashed forward, space and time began. Our universe expanded out from that single point in time and space. The universe expanded, energy became light, and hydrogen and helium were created. Matter formed as the universe continued to expand. Worlds developed and life formed.

Our science currently tells us that energy can neither be created nor destroyed. The energy in our world, the energy in you and me, the energy in all the people in this airport and in every employee working in every business on the planet was present at the beginning. Think about that, the creative energy that started at the beginning of creation and made each and every thing that we know is contained in all your employees.

From the moment of the Big Bang the universe has never stopped creating. Just stop, take a deep breath, and look around you at the marvelous things created from the minds of men and women. I just want to repeat again. Go back just 100 years and look around, then pop back and wonder where all the changes came from. You have only to realize that this creative energy is still flowing through every single human being just waiting to be developed. What a wonderful world we could have as we unleash this creative power locked inside us all.

Following is some information regarding our ever changing universe from a May 19, 2004 *New York Times* article by Dennis Overbye, called *"By X-Raying Galaxies, Researchers Offer New Evidence of Rapidly Expanding Universe":*

1.   "Observations of giant clouds of galaxies far out in space and time have revealed new evidence that some mysterious force began to push the cosmos apart six billion years ago, astronomers said yesterday."

2.   "The universe is accelerating," said Dr. Steve Allen of Cambridge University in England. "We have found strong evidence for dark energy."

3.   Dr. Michael Turner of the University of Chicago, said: "We can now be quite confident that the expansion of the universe is speeding up. It's not a fluke, it's not going away."

4. "Clusters of galaxies are the largest objects in the universe, containing thousands of galaxies and trillions of stars. But in a big cluster, the stars themselves are greatly outweighed by intergalactic gas, which has been condensed and heated to 100 million degrees or so by the cluster's immense gravity."

A story comes to mind. A consultant asked a supervisor to point out his least motivated worker. "Easy. It's Harry." "Okay, I want you to study Harry this week and find something great about him." "Impossible," said the supervisor. "If it wasn't for the union, Harry would have been gone years ago. There is nothing good in Harry." The consultant insisted. So, the supervisor followed Harry around that week, even after work. One day he saw Harry hit a home run and hustle around the baseball field. He couldn't believe it was the same Harry that worked for him.

When the consultant came back, the supervisor was bubbling with enthusiasm that he had completed his assignment and found something positive about Harry. With his new eyes he was able to see something different within Harry and was able to relate to him more respectfully.

This story demonstrates that we should realize that people must be careful about their own perceptions and misperceptions about others people and that people will often respond in the way we think about them. With new objectivity, you can see your employees with new eyes and can draw out new ideas from them.

Begin by recognizing that every person has creative ability and that with a little patience and support, you can nurture that creativity. When you start to bring the

ALL YOU GOTTA TO DO IS ASK

suggestion box to the workers and simply ask them for their ideas, you will be amazed at what will come out of them.

Chapter VIII

Creativity Is Unending

"Dissatisfaction is the mother of improvement."—Shigeo Shingo

When faced with a problem, try to believe that there is always a creative solution. The most difficult thing is to recognize that you have a problem, for once you can identify a problem, the solution will always come to you, in time.

The jewels of creativity are lying inside every employee in your company. You just have to go to your people, look at them with new respect, and ask them to help your company be more successful from their ideas. You tell them that you need them.

It is a win-win process for all. When employees create new ideas and the ideas are implemented, they will feel really good about themselves. You will feel much better about them, as you can see within them a much deeper talent then you did before. In addition, you will feel better about yourself, as you were the leader of the process. The company will also benefit because it is making more money from the ideas that result.

It's important for managers and supervisors to realize that taking the suggestion box to employees each and every day creates a structure for communicating within the organization.

I hope you realize that we are not saying that you have to actually lug a big metal suggestion box around the factory or the office. No, we are using the expression figuratively. We want you to actively go to the workers each day and ask them for their improvement ideas with a smile (not with a physical box).

Remember, the process is this: the immediate supervisor, middle managers, and others go to each employee and learn directly from the employee about his or her ideas. This allows problems to be solved where they are encountered. The employee and supervisor implement the employee's idea where the employee works.

As mentioned earlier, Dana Corporation, a large automotive supplier, started years ago to bring the suggestion box to its employees and now the company receives two improvement ideas per month per employee with 80 percent of them implemented. Fidelity Investments also does it, Toyota does it in America, and it is starting to take hold here.

WHAT SETS US APART?

"The Toyota Production System is at the heart of everything we do. Based on the concept of continuous improvement, or kaizen, every Toyota team member is empowered with the ability to improve their work environment. This

includes everything from quality and safety to the environment and productivity. Improvements and suggestions by team members are the cornerstone of Toyota's success."

Let's take a brief look at how this started at Toyota.

Eiji Toyoda and Shoichi Saito of Toyota came to the United States to study automobile factories. They noticed workers making suggestions for organizational and production improvements under the Ford suggestion program at Ford's Rouge complex in Dearborn, Michigan. In 1951, they started Soui Kufu Teian Seido (ideas suggestion system) at Toyota.

Also, in 1951, four instructors from the United States began training in Japan to develop trainers for Training Within Industry (TWI). This training was originally created by the U.S. government to increase the U.S. industrial productivity and quality to support the war effort during World War II. TWI included training supervisors to teach workers how to make changes in their jobs. Jobs Method Training, a part of TWI, focused on modifications for improvement. The Japanese Labor Ministry continues to promote TWI. Many companies in Japan have their own in-house Job Methods Training for managers and supervisors.

Employee suggestions combined with Jobs Methods Training was quickly recognized by Toyota as a competetive advantage.

All ideas are encouraged and, of course, not all ideas are successful. Taiichi Ohno in his book, *Workforce Management,* said this:

ALL YOU GOTTA TO DO IS ASK

"We should be happy if half of our instructions or orders are correct. In olden times, Confucius surely knew that we were wrong half the time . . .

"When allowing people to try out an idea, the person who gave the instructions in the first place should be present to follow the results closely. If the idea does turn out to be a mistake, then the fact that the error is witnessed firsthand will have an effect on the workers. They will realize that since the boss apologizes when he or she makes a mistake, they as workers can feel freer to experiment with whatever ideas might occur to them.

"On the other hand, wouldn't workers be even more cooperative when mistakes are met, not with reproving looks, but with encouragement and the explicit recognition that only five out of 10 ideas that you yourself come up with are right? When workers start thinking that they have to keep quiet and stick with whatever the boss tells them to do, for better or worse, they will gradually stop listening."

Supervisors became champions of the idea implementation system at Toyota.

When trusted to do so, workers can resolve problems caused by actions under their control. Management can allocate resources and provide the tools, training, and environment necessary to ensure the quality of the ideas implemented.

To make the system of taking the suggestion box to the worker work, it only takes your leader, your CEO, your plant or office manager to ask everyone to offer two improvement ideas per month to make their work easier and more interesting. Then you need just one person in the company to become the bandleader, to continually inspire people to come up with new ideas to sustain the process.

He or she will become the driver, steering the process to drive down costs and drive up quality. One person is all it takes to provide the discipline for sustaining the ongoing effort. Multiplied by all managers and supervisors, imagine the results.

In conversations with dozens of people across all levels of management from senior officers to front line supervisors, the following themes are always included when asked, "What is the role of supervision / management?"

- Stimulate interest;
- Provide resources;
- Train;
- Coach;
- Feedback;
- Encourage.

The supervisor's role is similar to that of a sports team coach. His or her primary focus is to help develop the skills of the team members and to provide the necessary resources for them. But too often, supervisors feel their role instead is to:

- Watch. Make sure that people are giving a fair day's work for a fair day's pay.
- Control. See that people do not steal or try to beat the system.
- Record. Keep the payroll records, necessary of course, but not the prime focus of the job.
- Direct. Tell people what and how to do the job, not fully trusting them to do it on their own.

They use the old "carrot and stick" approach.

Tom Peters, in his book *In Search of Excellence* wrote, "Our general observation is that most managers know very little about the value of positive reinforcement. Many either appear not to value it, or consider it beneath them, undignified or not very macho. The evidence from the excellent companies strongly suggests that the managers who feel this way are doing themselves a great disservice. The excellent companies seem not only to know the value of positive reinforcement but how to manage it as well."

Supervisors should be the first to trust and respect their team members. And in truth, they are there to foster the creative spirit of every single person working with them. A good supervisor gets "work done through other people." The supervisor should go and listen to the workers and ask them for their ideas on how to improve the process and deliver better products and services to the customer.

People continue to walk by, some walking slow, while others seem to dart around, weaving in and out between others. Sitting in the airport and watching these people go by sometimes causes a chuckle and recalls other memories, like that Dr. Seuss story with a cat in a tall red and white stripped hat doing some pretty funny things.

Chapter IX

Humor

"If stupidity got us into this mess,
then why can't it get us out?"—Will Rogers

A couple of tools are available to every supervisor, which should be incorporated into the environment: they are humor and a smile. Humor helps to create a bond between two people. While our work must be taken seriously, we can make light of our situation. An organization can be too stuffy or serious, creating a structure that stifles interaction and inhibits creativity. A smile offers benefits for both the supervisor and the employee. A smile on the supervisor's face becomes contagious. It's good for morale.

Research shows that a person can increase the amount of a chemical called serotonin in the brain by changing his or her facial expression. Higher amounts of serotonin make us feel better, reduced amounts cause us to feel depressed. The tools of a smile and humor can be especially effective, good for our health, and they fit within every organization's budget.

According to Dr. Peter Senge, author of *The Fifth Discipline, The Art and Practice of the Learning Organization,* "Our organizations work the way they work, ultimately, because of how we think and how we interact. Only by changing how we think can we change deeply embedded policies and practices. Only by changing how we interact can shared visions, shared understandings, and new capacities for coordinated action be established."

So go out and ask your employees, "What can you do to make your job easier, more interesting, build your skills, and help the company save some money, improve safety, reduce defects, improve customer service, and reduce the time it takes us to deliver our products and our services?"

And then, listen with a smile. And never, never criticize the employee's ideas; never. Why? Because when you criticize, you will close them up like clams and they will never offer ideas again.

If you don't like the idea, you can say, "Thank you for your idea. Maybe you can slightly change it this way, or discuss it with your team members and see if they can help you implement the idea." People accept criticism much more easily from their peers than from their bosses.

Set the Tone

When new ways are found to cut cost, reduce delivery time, and increase quality, the customer still wants it better, faster, cheaper. The search for continuous improvement goes on.

Senior managers know that to satisfy customers (as well as owners/shareholders) the various parts of a company need to be more efficient and more responsive. The best managers and supervisors share one common characteristic—**they all produce better results quicker.** Not just today and not just tomorrow, but the day after and the day after that. What is needed is a process for continuously driving down costs while continuously increasing service.

At times senior managers seek new ways to cut costs as well as better service customers through groups of selected employees. These groups are used to develop valuable ideas. Most often the process is sporadic; it happens only sometimes, not consistently, and it does not involve everyone.

Product developers are asked for more innovative and creative new products and services. "We need great products that truly excite our customers." Leaders talk about change—the need for change, how to change, and why change is important. And leaders do get cost reductions and companies find some new ways to satisfy the customer. However, they do not often enlist everyone in the search on a continuous basis, because as soon as some improvements are made, they stop looking and move to other priorities.

ALL YOU GOTTA TO DO IS ASK

Chapter X

Listening to Customers

"If the customer doesn't vote with a favorable decision, nothing else matters."
—Gerald A. Michaelson, Sun Tzu: The Art of War for Managers

We have been talking about going to the employee for their ideas and then listening. We should also learn how to listen to our customers. I noticed lately that whenever I have a problem and call customer service, the first reaction I get is something defensive, putting the fault back on me. I would say this happens 90 percent of the time. "You didn't read the instructions carefully enough!" "Our guarantee was for a limited life, not for that kind of wear." And the one I like the most was, "I didn't make a mistake. Someone else in the company did."

Whenever my tone is accusatory, people react defensively.

I once called the customer service manager at Stew Leonard, a giant grocery store in Westport, Connecticut. They have carved on a large stone in front of the store, "Our Policy: Rule 1 - the customer is always right!

Rule 2—if the customer is ever wrong reread rule 1."

I asked the manager at Stew Leonard, "How is it possible that the customer is always right?" She said, "When the customer calls to complain, I listen and let the customer talk it all out. I never react immediately. Then when the customer is calmer, and they do appreciate when you just let them speak, then you look for a way that they can always be right."

It is an art and only needs a little practice.

Respecting Your Associates

Once, while I was working with a data processing company, the company totally ran out of work. There were 190 data entry operators with no work. A quick meeting was held just before closing on a Friday night and all of the employees were asked to talk to their friends over the weekend and to inform them what the company does.

On Monday morning a systems analyst burst into the office excitedly. He said he had gone to a party on Saturday night, met a manager at The Bank of New York, told him about the company's services, and found out that the bank needed exactly what our data processing company did.

A few days later, our company received enough work to keep all of the 190 operators busy for the next month. Amazing things can happen when you ask your employees for help.

"FAILURE TO CHANGE IS A VICE!"

Hiroshi Okuda, Chairman of Toyota said, "Failure to change is a vice." He went on to say, "At the very least, you should not be an obstacle for someone else wanting to change."

Imagine. Toyota, probably the richest most successful company in the world, creators of lean manufacturing, is asking all of its employees to change. Toyota is incredibly successful as is. Its market capitalization at the time of this writing is worth more than General Motors, Ford, and Daimler Chrysler combined. You would not think that Toyota would need to change. Yet, it seems that change is one of the company's secrets to success.

ALL YOU GOTTA TO DO IS ASK

Chapter XI

Unrealistic Expectations

"But the fact that some geniuses were laughed at does not imply that all who are laughed at are geniuses. They laughed at Columbus, they laughed at Fulton, they laughed at the Wright brothers. But they also laughed at Bozo the Clown."—Carl Sagan

The other day I spoke with a technical group supervisor. I was visiting a manufacturing plant and was interested in finding out a little more about an employee's improvement idea. A few months earlier, a production worker had come up with a new idea of drilling two extra holes in a squeegee, allowing machine operators to turn the squeegee around when one side wore out, allowing the other side to be used. This was a cost saving idea that would cut squeegee costs by 50 percent per year.

Earlier, I had talked with the employee who showed me the squeegee and how it fit into the machine. He demonstrated how two additional holes allowed doubling the usefulness of the squeegee. Afterward, when I spoke with the technical supervisor to ask if someone from his staff assisted in drilling the holes or testing the idea, he was shocked. He proceeded to tell me how turning the squeegee wouldn't work. Integrity of the machine would be compromised, quality wouldn't be acceptable, etc, etc, etc. He said the idea was tested over a year ago with bad results.

I was confused. The idea was supposed to have been implemented months prior. I had recently spoken with the employee who submitted the idea and he showed me how it was done. I decided to do a little investigating. I asked a machine operator to show me where the squeegees were stocked. All the squeegees had the two additional holes. I asked if they were being turned around and reused. The answer was yes. I asked another operator and received the same answer, and that the machine worked fine and there was no reduction in quality. I asked another and another. Each time I received the same answer.

People started telling me that the technical supervisor didn't accept their ideas. Production workers didn't know about machinery; only technicians could suggest improvement ideas. This supervisor was stuck in his belief that a change couldn't be implemented unless the idea came through his technical group. He didn't change his thinking despite the fact that many improvements were implemented and were now 'standard operating procedure' in the plant. He failed to see the change occurring around him. His mental model of how the plant operated was much different than the current reality.

Dr. Peter Senge said, "Mental models are deeply held internal images of how the world works, images that limit us to familiar ways of thinking and acting. Very often, we are not consciously aware of our mental models or the effects they have on our behavior."

The mental model that a change couldn't be implemented except through the technical department did not allow the supervisor to recognize that the machine operators were, in fact, turning around the squeegee and saving the company half of the money previously spent. Even worse, this mental model made him discourage ideas suggested by production workers. His staff knew this, so they didn't support improvements "coming from the floor." The production workers knew this, and it took the operations supervisors a lot of effort to try to overcome this obstacle. Sometimes they succeeded (as in the squeegee idea) and sometimes they didn't. Many supervisors wouldn't fight for their worker's ideas; management viewed the technical supervisor as something of a hero, working hard to fix machinery when it broke down. (It makes one wonder about the effectiveness of maintenance programs.)

The organization may never know how many improvements weren't made because people felt their ideas wouldn't be accepted.

Here's another story about unrealistic expectations. I recently had dinner with the general manager of a car dealership in Ohio. His dealership was part of a group with multiple locations. He told me how hard he was working; long days, sometimes seven days a week. During the course of the conversation he told me how he started in sales and progressed to his current position. He told me that in the car business, that was typical. Salespeople, if they were good, could sometimes become sales managers and then general mangers. In fact he didn't know of any general manager who followed another path, not service or finance, always sales.

He remarked that the only thing that mattered was

sales. That was how he was always measured and that is how he is currently measured. Sales are the most important thing in a car dealership. He stressed this point time and time again. He said that everyone in the industry knew it, that is how people are measured, and that people in other businesses often don't understand car sales.

Interestingly, when I asked him how much time he spent developing his sales people, he said he didn't have time to devote to them. He would show them a two-hour video about the product and turn them loose. If they weren't any good, he would let them go after 90 days.

As we talked, it became apparent that he knew the car business and he thought he had all the answers. However, he did have a problem. Sometimes the sales process took too long and he would like to shorten it. He had hired a high-priced sales consultant to come in to speak with his salespeople, but it didn't help. He figured that even though he didn't yet know how, he was sure that technology was the answer.

I was intrigued. Here he was, a man who knew all about the business. The person with all the answers. Yet his sales weren't meeting organizational expectations. I knew that if he took the suggestion box to his people he would get some insights. At the very least, he would realize that he needed to drive the sales effort to help others better themselves. I wondered if he would ever realize this.

In his view, sales are the most important thing. It is the only goal. Yet during the long hours he worked, this manager devoted no time to ensuring the success of his salespeople.

Contrast this to another business, Fifth Third

Bancorp, based in Ohio. There are some similarities between banks and car dealers. Both grow through sales and both provide services. They have a lot of competition, some with the same products and services and some with slightly different products and services.

Sometimes a product is a differentiator and a customer will be attracted by something unique in the marketplace. More often, however, customers are attracted by how they are treated as human beings. This human service element is important in both car dealerships and banks. Fifth Third Bank understands this and being able to execute it on a day-to-day basis becomes a competitive advantage for them. At Fifth Third Bank, everyone is a salesperson. They have programs and incentives where every employee at every level can earn a commission for referring customers to the bank's products.

I don't believe our car dealership manager understood this kind of service. He thought service meant servicing a vehicle.

Bringing the suggestion box to the employees, and asking them about their customers and workflow problems can highlight areas where service can be improved. People can also point to waste that leads to lost opportunities or profit. I know people who have visited car dealers, waited for 10 minutes to speak with a salesperson, then left when no salesperson could be found. Did the general manager know how often this occurred in his dealership? Did the receptionist know? Was there a plan for someone to start interacting with the customer when a salesperson wasn't available? Would the receptionist at the car dealership receive a commission for starting the sales process with a customer? Probably not, and the receptionists likely do not see themselves involved in sales. At best, they page "New car sales" or "customer waiting" over the intercom.

This human service element is extremely important in maintaining and satisfying customers. This is true in many businesses.

Another example is a physician's office. Unless they are victims of malpractice, most people are unable to judge the competency of their doctors. People can and do judge their doctors based on the human service element. How long did they have to wait? How clear were instructions? Was the office neat and orderly? Did the receptionist smile?

During dinner with the general manager of the car dealership, each time I tried to bring up the creative potential of involving employees and bringing the suggestion box to his workers. He stated, "You know the car business. It is different than all other businesses, and the only thing that matters is selling a car."

The sad ending to this story is that we agreed to continue this conversation the following week. I collected some information about car sales, customer service, and lost opportunities, and hoped for a productive dialog. Unfortunately, a couple of days prior to the meeting, I found out he was no longer with the organization.

There are always other priorities that get in the way; urgent customer requests, safety problems, environmental issues, personnel problems, and other fires that have to be put out. How will we deliver what the customer wants? How does the competition do it? Is our marketing working? How can we measure its impact? Are we supporting sales? Are our finances in order? Are we recruiting the right people? Is our training effective? These questions often take over. We continue to measure things, such as process capabilities, financial performance, and delivery performance.

And what about those frustrated managers, supervisors, and unhappy workers? They are often ignored. Although we just cut costs, reduced delivery time, and increased quality, the customer still wants it better, faster, and cheaper. We ask for more innovation and creativity in new products and services. We talk some more about change; the need for change, how to change, and why the change is important. But we never realize the potential of our employees in finding the most effective ways to change.

> *We watch people walking in the airport, some with luggage, others with family or business associates. Some casually stroll while others hustle to a gate or ticket counter. We have goals and many times these goals are independent of other goals within the organization. Although at times we act independently, we are all participants in the process and we have concerns. In the airport, the concerns may be simply to find a decent meal or something more important such as missing a flight.*

ALL YOU GOTTA TO DO IS ASK

Chapter XII

Here Is How It Works

"We are surrounded by insurmountable opportunity."—Pogo, Walt Kelly

Those frustrated managers and supervisors and unhappy workers have concerns. The lack of resources—who will do the work? The lack of time—it takes time to make a change. Changes run the risk of more bureaucracy and may become too invasive. Some managers fear that their people will resist change. Workers feel their job security is in jeopardy.

Let's not lose track of what we truly want to achieve, to become more competitive while enabling us to become more responsive on an ongoing basis.

Who has the answers? If we are lucky and we have some bright leaders, then they might discover all by themselves how to bring change, excite and motivate workers, and please our customers. **Wise leaders, however, understand that success means getting everyone involved.** It is as simple as bringing the suggestion box to all of the workers. The adage that "two heads are better than one" is an understatement; 500 heads are much better then a dozen or two. So let us challenge everyone in the company to try to come up with ideas on how to make things better. The secret is to look for small ideas, and to look for lots of them.

Here is how it works:

1. Challenge all employees to come up with at least two small ideas each month to make their work easier or more interesting, reduce costs, improve quality, improve the throughput, improve safety, or to improve customer service;

2. Encourage employees to write ideas down on an idea form every time they find a problem, make a mistake, or see an opportunity for improvement;

The Idea Form

Quick & Easy KAIZEN	
Before Improvement	After Improvement
The Effect:	
Date	Name

3. When employees get an idea on how to solve a problem, have them write it down, along with the effect and the benefits when implemented;

4. Have employees discuss their problems and ideas with their supervisors and fellow associates. (Note to supervisor: Never, never criticize the idea—all ideas are great. They all might not be implemented but one idea leads to other ideas if you don't shoot them down.) All ideas submitted must be followed up within 24 hours. Managers and supervisors must get back to the workers to let them know that their ideas are valued;

5. The person who comes up with the original idea should implement the idea him/herself, or with his/her work team. The person who comes up with the idea should stay in charge of getting the idea implemented, even when help is needed from other departments, i.e, maintenance, engineering, software, or accounting.

6. If the idea is implemented, the completed form is submitted to either a supervisor or person in charge of the idea system. If the idea is not implemented, then the supervisor or the person in charge can offer alternative suggestions to the employee or go after additional resources to help get the idea implemented;

7. Implemented ideas may be recorded in an idea log. Unimplemented ideas submitted by employees could also be recorded;

8. Implemented ideas should be posted onto bulletin

boards displayed for everyone to see. This encourages workers to admire ideas, to learn from each other, and to copy those ideas to make their work areas better. Unimplemented ideas could also be posted separately on bulletin boards to inspire others to help bring them to completion and to ask help to get them implemented. Both implemented and unimplemented ideas can inspire others to come up with new ideas on their own.

Whenever possible, take or draw pictures of the *before* and *after*. Viewing a picture is easier than reading words, and it is exciting to look at. A picture quickly explains the improvement idea and stimulates other people to do the same thing.

And yes, we do want people to copy each other. We are looking for continuous improvement and we want people involved. We can learn a lot from others when we copy their ideas—as we copy, we often come up with new ideas to make the original even better;

9. Keep monthly statistics to ensure that the goal of two ideas per person per month is reached.

In conversations with dozens of managers/supervisors, the following themes are constantly repeated when asked, "What is the goal of the employee involvement process?"

- To make work easier;
- To give people some control over their work environment;
- To develop people to learn new skills and to have greater awareness of the problems and potential problems around them;

- To develop problem-solving skills;
- To save money;
- To improve quality;
- To unleash creativity;
- To improve safety;
- To enhance communication;
- To better customer service;
- To improve work process flow;
- To increase involvement in the company;
- To make people's jobs more interesting;
- To improve the throughput.

By giving employees simple idea forms to complete, managers are essentially taking the suggestion box to the employee. They learn directly from the employee about new ideas. The supervisor and the employee are communicating, working together, and implementing the employee's idea where the employee works. The workplace is improved.

Quick & Easy KAIZEN	
Before Improvement	After Improvement
Wires are hanging and laying on the floor behind the ship station.	I tied the wires together and put them on a hook out of the way.
The Effect: When cleaning or adjusting the wires will not trip anyone or get in the way	
Date 5/1/04	Name Mary O Reilly

Which Pleases You More?

Which would you like to do, implement your own idea or an idea that comes from your boss, your wife, your mother, or even your mother-in-law? Most people can get very excited about coming up with an idea and implementing it rather than to work on someone else's idea. It is human nature. Even though this is true, we still want people to copy other people's good ideas to make the work environment better. We want to have everyone involved and some people need help to get started.

Captain Jean-Luc Picard (played by Patrick Stewart on *Star Trek: The Next Generation*) of the Starship Enterprise responds to ideas by saying, "Make it so." He understands the power of having his people implement their own ideas.

Supervisors now have a perfect opportunity to set the tone and ask for ideas that will lead to reducing costs and increasing service. Since they are communicating with employees on a regular basis, it takes minimum effort to do this on an ongoing basis.

The supervisor can ask for concrete ideas on a specific topic such as improving quality, safety, throughput, customer service, or reducing costs. Then they can share the ideas they've heard to feed the ideas and present problems to workers on a regular basis in a safe, non-threatening manner.

Supervisors and managers should walk their work areas daily and point out possible opportunities where the company needs help in solving problems.

Some workers will never submit an idea. But you will find that almost all of the employees will participate over time. Once they overcome their fear of being criticized, slowly you will see wonderful ideas emerge.

Jim O'Dell, Vice President of Operations & Systems for the Tribune Publishing Company in Chicago, Illinois, stated during an interview in _New Horizons_ magazine, "In the course of just a few years, the [Chicago Tribune's Innovator Suggestion] program went from generating 16 ideas in a year to more than 1,000 ideas each year, and with cost savings of $3 million annually. Of course the program brought about process improvements, quality improvements, cycle time reduction, a reduction in operating expenses, and improvement in customer service.

"But just as important, there was a higher level of morale. Employees began to feel like owners of the company because they were involved in the decision-making and the success of the company. One thing we learned was to reward small victories. We didn't look for home runs; we looked for lots of singles. You'll get a home run every now and then, but you'll find a lot more people capable of hitting singles."

When you bring the suggestion box to the workers and ask them for their ideas, you will be amazed at the excitement and the positive change that starts to take place.

Your reaction will be, **"Why in the world haven't we done this before?"**

Somehow we just get stuck with misperceptions about the people that work for us. Companies design a "lousy job," give it to workers, and they do it without complaints; and employers think, "They must not be too smart to do that same job every single day!" Yes, the workers do conform to your image of them. It reminds me of a story from General Electric (GE) about people living up to (or down to) their supervisors' expectations.

A consultant asked a supervisor at GE, "Tom, of the ten people that work for you, who are the highly motivated people and who are the hangers-on?" Tom wrote down five names and put them into each category, highly motivate ed and hangers-on. Tom then was transferred to another group. A new supervisor was put in charge and was told that she had five highly motivated employees and five barely doing their job. At the end of one year the same consultant asked the new supervisor to rank her employees. She ranked the ten exactly as she was originally told, the complete reverse from the previous supervisor. Yes, people often conform to the manager's previous set images about them. But, if the manager looks at them differently, they respond differently. Isn't that just amazing?

It's critical that all people at every level in the organization know that their ideas are welcome. It's important to let people know that all ideas, even bad ideas, are welcome. It's tough for people to offer ideas, especially good ideas, if they are not aware of problems or challenges. Supervisors need to let their workers know what's important, from absenteeism to customer service. Supervisors must work to make employee creativity part of the corporate culture and actively solicit ideas.

In the past, the supervisor was often the obstacle to new ideas from employees because the supervisor felt it was his/her job to come up with ideas. I have heard several cases where supervisors actually stole ideas from their employees. However, when senior managers ask for two ideas per worker per month, the supervisors realize that they must look at their employees differently. And when they ask for ideas, they get them.

In 1984, Tôhoku Oki Electric Co. of Fukushima led all of Japan for the second consecutive year with an average of 570 suggestions per employee.[5] Sounds crazy! What was their secret? Did it really work? I had to find out.

I was curious about Oki Electric's results, and I doubted it worked. In 1985 on one of my study missions to Japan, I took a group of American managers to visit Oki Electric. In the lunchroom and in the meeting rooms, I saw very large notebooks with thousands of employee suggestions. I couldn't tell what the suggestions were, because I did not read Japanese. I asked the senior manager at Oki Electric to introduce me to the person who came up with the most ideas in the past year. He did. Proudly, the idea leader took me to room in the plant to show me his job. There was a machine that automatically produced parts without any operator at all. With his ideas, he had discovered a way to perform his job without him.

Every night he would go home with his head filled with the problems to be solved. With his wife's assistance, he wrote down the problems and the small steps needed to solve them.

It sounds crazy implementing close to 15 ideas per day, but I saw it with my own eyes. He knew how to break down his ideas into very simple improvement steps:

1. Turn the tape one degree to the left;
2. Sand the surface smoothly;
3. Adhere the tape to the surface;
4. Etc.

ALL YOU GOTTA TO DO IS ASK

Each step is a very small improvement and the net result is great cost savings to the company. I am not saying that you have to follow Oki's ways. Perhaps, Tôhoku Oki Electric is the exception, not the rule. But everything I read, every study I've looked at, shows that the average Japanese production worker makes somewhere between 20 to 24 suggestions per year, while their counterparts in the United States make less than one. This will change soon as we realize the potential benefits that come from getting everyone involved in improvement activities.

In fact, even if you can save either one penny or save one second each day, it is an excellent idea. Pennies and seconds add up. What you want is continuous improvement and you want everyone involved.

At the end of the year 1984, Oki Electric had spent 70 million yen ($304,347) on improvement ideas and saved 1.09 billion yen ($4,739,103). That represented a pretty nice return on their investment.

Mamoru Kanno of Tôhoku Oki's Machining Section 1, First Manufacturing Department, submitted a whopping 3,226 ideas that were implemented in 1984. Shizuo Takahashi from Assembly Section 3, Second Manufacturing Department, submitted 3,046 ideas that year. (*Japan Economic Journal*, October 26, 1985.)

Oki Electric had 817 employees at the time. The second closest company, Taiho Kogyo Co., averaged 220 suggestions per employee. In Japan, the average was 24 ideas per employee per year, the same as Dana Corporation in America. Even though it might sound crazy, it works. It definitely works and it works very well.

Bringing the suggestion box to the worker gets

every employee in the company thinking about and doing something about continuous improvement.

I spoke with David Harman, Manager of Continuous Improvement for ILC Dover, and asked him, "What's in it for the worker?"

David Harman, Manager of Continuous Improvement for ILC Dover, was asked, "What's in it for the workers?" Harman replied, "From the employee's point of view, it gives them hope and a measure of control over their job. It provides them an avenue for expressing what they may otherwise suppress or vent in unconstructive or uncomplimentary ways. The greatest danger, of course, is to implement a program and then make it a charade or hollow."

ALL YOU GOTTA TO DO IS ASK

Chapter XIII

The Manager's Walk

*"I couldn't wait for success so I went
ahead without it."—Jonathan Winters*

Companies throughout Japan use a practice called
the "manager's walk." It is a major factor in leading
improvement efforts.

I was introduced to the manager's walk in 1981
when I visited a Sumitomo Electric plant in Japan. It was
11:00 a.m. when the manager invited me on his daily walk.
The plant manager stated, "I do this twice a day, every day,
and I feel that this is the most important part of my job."

Plant managers have a walk through the plant at the
same time each day. A different theme such as "waste of
motion" or "causes of employee turnover" is selected for
each walk. The plant manager will meet with each
department manager/supervisor, ask questions, and share
information learned. For example, if the first supervisor in
the first department has a method of reducing wasted

85

motion, the manager will listen and discuss the idea. Then the manager walks to the next department, will ask the same question, and will share what he learned earlier. This is a very powerful process of learning and sharing information.

The Keys:

1.　　Power exists at the top. The manager has the budget and the power to bring change to the facility. When a manager considers something important, others will support it. When a manager neglects something, there is a tendency for others to also ignore it;

2.　　The manager asks questions to learn. Questions should be prepared in advance but allow for spontaneity. Rather than use the questioning method often used by teachers, where they look for answers they already know, the managers asks questions for which the answers are not already known;

- *How do we use quality charts?*
- *Why are they important?*
- *Do the employees look at them?*
- *Are they kept up to date?*
- *Who keeps up the charts?*
- *Are there any problems with them?*
- *What do the employees think about the idea system?*
- *What are the problems?*
- *How can we improve the system?*
- *What percent of the employees submit written suggestions?*
- *How can we get more ideas implemented?*

3.   The manager listens carefully. This is not a dialog. The manager wants to learn, and he/she learns by listening, by gathering information. The manager listens carefully, not judgmentally. He/she allows the presenter to be the expert, praises good ideas and then to stimulate possible areas of improvement may ask questions such as:

- *What are our goals for this theme?*
- *How do we visually display the goals?*
- *Who is responsible for maintaining them?*
- *How often are they updated?*

4.   The supervisor is then encouraged to walk his or her area to ask the employees questions on the same area of concern and to share information with them on the theme;

5.   At the end of walk, the manager should summarize and share the ideas learned with all of the employees by posting the information gathered on a wall or in a newsletter. The posting should include:

- *What was learned;*
- *What areas can be improved;*
- *What things can be praised;*
- *What follow-up efforts should be made;*
- *Time and date for follow-up items to be completed;*
- *How they will be communicated.*

On the day of my first walk with the plant manager at the Sumitomo Electric plant, we looked at quality control charts. The plant manager wanted to ensure that they were being used and updated and that were adding value. When the plant manager shows interest in what is being done, the supervisors and employees will support it by using the charts and keeping them up to date. When the plant manager shows no interest, in most cases, the charts will not be used and kept up to date. I have been to many plants where charts are shown, but are not kept up, most workers ignore them, and they lose their original meaning.

Selecting one topic or area of concern will help employees generate specific ideas rather than generalities. Create a separate theme for each walk, such as:

- *Safety ideas;*
- *Cost savings;*
- *Better customer service;*
- *Better communications;*
- *A specific machine;*
- *Lighting;*
- *Smells;*
- *Temperature;*
- *How things are stored;*
- *Transportation;*
- *Health;*
- *Air quality;*
- *Potential hazards;*
- *Repetitive strain injuries;*
- *Telephone procedures;*
- *Wasted paper;*
- *Extra inventory;*
- *Movements and motion;*
- *A better way to do things;*

The topics are unending!

The manager becomes the walking suggestion box by going where employees work and asking for help. The manager has given employees permission to make recommendations. It is important to let employees know that they will be implementing most of the ideas. It is essential for the manager to teach the employees how to quantify their ideas in terms of time saved, material saved, quality improved, rejects reduced, and customer satisfaction.

Imagine what happens when you pick a different theme every day for your walk. At the end of the year, you will have covered over 220 different topics.

The manager's walk allows you to specifically identify problems for the employees. During the walk, a supervisor needs only to mention, "We're not meeting our on-time delivery schedule" or something similar related to downtime, throughput targets, or departmental goals. This allows an employee to become aware of problems or concerns.

These questions made during the walk get people thinking about improvements. Ask questions such as, "How can we reduce inventory in your area?" or "How can we reduce the set-up time on this machine?" or "Is this machine a problem for you?" These type of questions trigger ideas.

Being a walking suggestion box is more than an exercise or an activity. It needs to become a way of life for management.

The Gemba Manager's Walk

Purpose and Definition of the Walk:

The purpose of the walk is to:

- *Strengthen leadership abilities;*
- *Gain alignment with business goal;*
- *Support better communications;*
- *Share information;*
- *Ensure that not a single detail is neglected.*

At least once a day, preferably twice, the senior manager will walk throughout his/her entire facility studying one specific topic on each walk that can lead to improvement.

Dr. Ohno's Seven Wastes:

1. **Defects** – Inspection, scrap, and repair;
2. **Overproduction** – Producing more than is required;
3. **Waiting** – Idle time;
4. **Transportation** – Transporting materials and parts more than absolutely necessary;
5. **Inventory** – Any supply in excess of one-piece flow;
6. **Motion** – Action that does not add value to the product or service;
7. **Extra Processing** – Effort that adds no additional value to the product or service.

Other Wastes:

Non-Utilized or Underutilized Talents
Underutilizing employees' potential. This is the mother of all wastes.

Workplace Organization (Five S's):

Use these five S's to help create themes:

1. **Sort** – What is not needed, sort through, and then remove. "When in doubt, throw it out!"
2. **Store** – What must be kept, make it visible and self-explanatory so everyone knows where everything goes. "A place for everything and everything in its place."
3. **Shine** – Clean everything that remains; equipment, tools, and workplace. "Cleaning is inspection."
4. **Standardize** – Implement best practices to keep the workplace clean and organized.
5. **Sustain** – Make a habit of properly using correct procedures.

Other Themes:

- *Quality;*
- *Safety;*
- *Cost reduction;*
- *Fire prevention;*
- *Flooring;*
- *Lubrication;*
- *Downtime of equipment;*
- *Training;*
- *Smells;*
- *Lighting;*
- *Standardization;*
- *Blockages;*
- *Storage;*
- *Repetitive strain injuries;*
- *Employee problems;*
- *Potential problems;*

ALL YOU GOTTA TO DO IS ASK

- *Packaging;*
- *Excessive movement;*
- *Change-overs;*
- *Visuals*
- *Noises.*

Chapter XIV

Improve Work Every Day

"You don't fly up a hill. You struggle slowly and painfully up a hill, and maybe, if you work very hard, you get to the top ahead of everybody else." —Lance Armstrong, <u>It's Not about the Bike: My Journey Back to Life</u>

For a number of reasons, there is a lot of resistance to implementing an involvement system for capturing employee ideas:

- A worker's idea is perceived as a threat to the supervisor;
- People are afraid of change;
- People are afraid of making mistakes;
- "I don't see any problems!";
- Coming up with ideas is the supervisor's job, not the worker's;
- "I have enough things to do. I could not handle a lot of new ideas from the workers;"
- "If they had any good ideas they would have been made a supervisor;"
- It is easier to complain than to do something

positive;

- "The employees are happy just doing their job and they don't really want to think about making improvements;"
- Who is going to implement the ideas?

Perhaps our culture is not ready for an idea system!

For some reason, the Japanese have been able to encourage their people to embrace ideas, but we in the West resist it. This can change as organizations accept this book and realize the infinite creative potential that can be tapped once people are asked for their ideas. The spark for the Toyota Production System, Just-In-Time, actually came from Henry Ford. James P. Womack cleverly changed the name of the Toyota Production System to Lean Manufacturing, the new name took hold and now it is sweeping America. It doesn't matter where something originates as long as our employees using the best tools and techniques to help our companies grow.

In talking with managers, even managers who have successfully implemented an idea system, concerns are heard with comments such as:

- "It's easy for a short while, but then ideas dry up."
- "We've picked the low hanging fruit."
- "Employees only have a limited scope and they won't (or can't) come up with many ideas."

At one company, where the goal is two implemented ideas per month, one of the most supportive managers voiced a concern that maybe six to 12 ideas per year could be achieved, but 24 was unrealistic. Perhaps 24 ideas per person each year takes work to achieve, but it is not unrealistic.

The following table shows what is possible when you have a creative Quick and Easy Kaizen system. (According to the HR Association in Japan in 2003.):

Ideas Per Employee in Japanese Companies			
Company	Ideas per employee	Number of employees	Economic benefit per employee
Aishin Light Metal	227.9	554	
Aishin Precision	17.7	6,295	
Aichi Machinery	82.3	2,141	$ 1,155
Aichi Seiko	36.6	2,090	1,863
Koito – head lamps	59	3,950	42,738
Gunze	48.8	4,734	929
Seiko Epson	35.9	10,235	8,570
Daido Metal	57	1,050	4,082
Daihatsu	19.3	9,118	2,226
Denso	12.4	28,482	
Toshiba Ceramics	57.9	804	249
Toshiba Ceramics	47.5	243	2450
Toyota Motors	9.2	58,000	
Nihon Hatsujo	61.5	2,167	6,632
Subaru	108.1	7,800	5,246
Matsushita	25.2	35,494	826
Yamaha	35.1	8,000	1,907

ALL YOU GOTTA TO DO IS ASK

Toyota used to get 50 to 70 ideas per worker per year but a few years ago, after 30 years of applying the system, they decided to shift from very small ideas to better ideas. This change may not have been a "good" idea for the original purpose of their suggestion system was to empower people to participate in problem solving activities through their own small creative ideas.

When looking at the list above, does it still appear that 24 implemented ideas per year is unrealistic? Of course, not all employees will participate at that level, and some may never contribute at all. But, some people may be very prolific. Remember we are looking for small ideas. In fact, the smaller the idea the better, the easier it is to implement. Each idea that is implemented benefits the employee and the company. Imagine how the workers feel when their idea is accepted and implemented. It is a wow!

Just look at the benefit Subaru obtained in 2003. With 7,800 employees and $5,246 economic benefit per employee they saved $40,918,800. Their investment in training and implementing those ideas was $70.05 per employee or $546,390, not a bad return on their investment.

Earlier in the book it was mentioned that ArvinMeritor an automotive parts supplier in Troy Michigan, had annual savings per employee of $4,285, at a cost of $204 per employee.

This is not about cultural differences between Japan and America, but about a different philosophy of management. This is about developing a culture of continuous improvement, and actively soliciting employee participation.

Here are some principles for nurturing this culture.

Daido Metal

Our Principles:

1. **We will always approach business from a global perspective.**
2. **We will always attend to the needs and desires of our customers.**
3. **We will continue to recognize the importance of the environment.**
4. **We will continue to value individuality and work to achieve synergy between people.**
5. **We will always remain open to new ideas and proposals.**
6. **We will always respond with speed and agility.**

Our Corporate Philosophy

1. **Our Duty:**
   We hold ourselves responsible for the happiness of everyone in our organization and the contributions we make to global society.
2. **Our Resolve:**
   We will create a vigorous and open-minded corporate culture through diligent self-discipline and ethical behavior.
3. **Our Foundation:**
   We will learn from our markets, respond to our customers' requirements, and surpass our customers' expectations.
4. **Our Approach:**
   We will devote ourselves to creation, innovation, and the realization of dreams.
5. **Our Objective:**
   We will strive to be the world leader in tribology[5] through constant technical improvement, development, and innovation.

"We, the **Toshiba Ceramics Group** companies,

---

[5] Tribology refers to friction, wear and the lubrication industry.

based on our commitment to people and to the future, and through our focus on the fields of highly-functional parts and materials and continual innovation, are dedicated to creating a higher quality of life for people, protecting the global environment, and to being a good corporate citizen contributing to the goals of society

Respect for Individuals
We respect every employee's individuality and creativity, and cultivate a dynamic corporate culture that emphasizes ability and achievement."

Chapter XV

The Principles:

Listen

"I do not know if you have ever examined how you listen, it doesn't matter to what, whether to a bird, to the wind in the leaves, to the rushing waters, or how you listen in a dialogue with yourself, to your conversation in various relationships with your intimate friends, your wife or husband . . .

"If we try to listen we find it extraordinarily difficult, because we are always projecting our opinions and ideas, our prejudices, our background, our inclinations, our impulses; when they dominate we hardly listen at all to what is being said . . .

"In that state there is no value at all. One listens and therefore learns, only in a state of attention, a state of silence, in which this whole background is in abeyance, is quiet; then, it seems to me, it is possible to communicate...real communication can only take place where there is silence."
—*Krishnamurti*

Listening is difficult. We naturally want to do something or say something. Listening, however, requires that we be quiet and confine our focus to what is being said. Sometimes leaders feel that to be good communicators they must be good speakers, but effective communication requires that they must also be good listeners.

Some bosses have a great talent in being able to ask a question and then answer it even before the employee has a chance to open their mouth. But we want the employee to answer the questions.

To release creativity in employees, managers must get involved in their employees' work. Involvement is demanding and requires listening. Some managers think that general praise, such as "good job" acts as a substitute for involvement. Workers know it doesn't and frequently the results are the opposite of what is intended.

Creative energy, like any other kind of energy, can be harnessed and managed. We've learned how to use the energy of the wind and the sun. We've learned to harness energy stored in the form of coal and oil. Entire industries have been built up around the discovery and management of sources of energy.

Discovering a source of energy is only the beginning. Managing and focusing energy allows us to get work done. Once we learned how to use heat to boil water and harness that steam to do work, steam engines began powering trains to transport people and material faster than horse-drawn vehicles.

Employees are creative, filled with unlimited energy

waiting to be released. Ask any supervisor about his or her people and have that supervisor tell you stories about how they get things done. Good workers can find new and better ways to do things. Others find ingenious ways to get out of doing work and creative reasons for being absent or late.

There is one energy source we are experts about. Unlike exploring for undiscovered sources of coal or oil, we know the location of this energy source: it's our workforce. Our challenge is not to discover the energy, but to harness and focus it. If creative suggestions are a good idea and employees won't come to the suggestion box, **then we have to take the suggestion box to the employees.** At first, this will be a way to harness their creative energy. Later it will be a way to focus the creativity within the workforce.

This resource, the creative energy in our people, is too important to ignore. Asking for simple improvement ideas is a start. Capturing ideas is like taking coal and oil and turning it into power. Roughly one-third of the energy in the fuel used in a combustion engine power plant is converted to electricity; the rest becomes waste heat and exhaust. In a car engine, only about one-fifth of the gasoline produces power, the rest is waste heat and exhaust.

Just because we can't turn 100 percent of the fuel energy into power doesn't mean we shouldn't use it. Initially only one-fifth of employee ideas may improve productivity. Isn't that better than not using the creative energy of our employees at all?

This creativity must be actively harnessed and focused on what really matters, which in business eventually translates to increased productivity and quality.

Actively managing the creative energy in the workforce means that we must understand how it is harnessed and how it can be focused.

Once we begin to get some ideas, we can share those ideas and find ways to build on those ideas to make gains. Turning the ideas that we would not initially implement into something useable is similar to the power plant that figures out how to take the waste heat and exhaust and create a cogeneration process using steam to drive turbines. Hybrid automobiles are another example of taking the unused energy, capturing it, and finding a way to use it.

We've discovered that we can take light energy, much of which exists around us in an unfocused form, and focus it. Light focused into a laser beam can be used to cut metal and transmit information, which can't be done with unfocused light. Focusing the creative energy in our people can bring the type of astonishing results that will never be realized with unfocused creativity.

Harnessing the creative energy in our workers requires a system to tap into the source (people) and capture the creative energy (ideas). In an organization, this means we need a system in place for this to occur. Someone needs to initiate or introduce the system into the organization. This may be someone at a high level; the CEO, president, managing partner, or perhaps even a human resource representative or a production worker. You could start in your own department and from your success let it spread throughout the organization. In any case, a champion or initiating sponsor with the power to sanction the new system and allocate resources (if required) is needed.

On an ongoing basis, other champions or sponsors are needed for the effort to manage this creative energy. These are the "walking suggestion boxes" who encourage and challenge people. Their tools are questions and sharing of information. At Toyota, these are the supervisors who take the responsibility to encourage those that have difficulty in participating. The supervisor also teaches the employees how to implement their ideas. It is the walking suggestion box who becomes the creativity management expert; harnessing ideas, focusing creativity, managing consequences, and providing educational opportunities.

Based on our experience, tapping into the creativity of the workforce creates much more than just improvement ideas. In discussions with managers and front line supervision many themes surfaced including:

- *Creates ownership in work;*
- *Changes the culture;*
- *Communication tool, fostering dialog between employee and supervisor;*
- *Puts brainpower of worker into the process;*
- *Having fun;*
- *Increases self worth;*
- *Creates synergy;*
- *Breaks down barriers;*
- *Builds trust;*
- *Helps people grow;*
- *Employees feel part of the company;*
- *Puts everyone on the same platform;*
- *Everyone contributes;*
- *Helps everyone get along better;*
- *Keeps everybody involved;*
- *Challenges people to be creative;*

ALL YOU GOTTA TO DO IS ASK

. . . and many others.

Chapter XVI

The Principles:

Cultivate

"Simple, clear purpose and principles give rise to complex, intelligent behavior. Complex rules and regulations give rise to simple, stupid behavior."—Dee Hock

People have a need to understand the role that their ideas have in the success of the business. People need to know the difference they can make by reducing variation, getting rid of defects, and trying new ideas.

As previously discussed, employees see many things their managers don't and they often know how to improve performance and reduce costs. Yet, they are rarely given a chance to do anything about it. It is time to encourage their involvement.

To cultivate ideas, managers must go where the work is done, observe, and consider the problems the employees face. The Japanese call this the "3 G's - Gemba (the actual site), Genbutsu (the actual thing), and Gensho (the actual phenomenon). Loosely translated it means come to the workplace, observe, and understand.

Pull the employees along by creating a vision. Set an expectation and ask them to help deliver that vision or expectation. Remember John F. Kennedy's vision of a man on the moon? Once he verbalized that vision, that vision was quickly realized. Employees will only have confidence in an idea implementation process if senior management supports it. Ideas have value only when they are implemented.

The walking suggestion box must be part of the normal operating process, not an add-on. People should know that their ideas are valuable, that their ideas will be acted upon, and many (although not all) will be implemented.

We have to create an environment where workers feel safe suggesting ideas both good and bad. To get the results we seek, we need a lot of ideas.

It is important to set a goal for the number of people you speak with on the manager's walk. Have some idea "seeds" with you at all times. Listen and share others' thoughts and ideas whenever possible.

Humor can be one of the most valuable tools a supervisor can use for cultivating ideas. Sometimes people think humor is not appropriate at work, but humor can help to relieve stress and help people find ways to enjoy work. People are naturally drawn to humor and want to engage in laughter. This is a good way for a supervisor to bond with their employees.

It's everyone's attitude that makes involvement

work. Encourage a culture where supervision sometimes catches people doing something right, rather than focusing only on catching mistakes. Make collection, stimulation, and implementation of good ideas a part of every manager's job and part of their performance evaluations.

Make idea implementation part of the supervisor's role. It should involve creating an environment that encourages ideas, helping employees develop their knowledge, and helping them improve their problem-solving skills to increase the quality and impact of their ideas.

Strangely, the best learning opportunities often come from the worst ideas. No matter how bad an idea seems, it should be treated as a teaching opportunity.

Many supervisors are surprised to learn the potential for performance improvement that lies at their fingertips, in their people's ideas. At Dana, 80 percent of cost savings comes from the employee's ideas. Supervisors should make certain their employees understand why ideas are important.

Train supervisors how to manage ideas. Teach them to listen, communicate, instruct, coach, and help their people develop. Supervisors must learn how to handle bad ideas, how to get initially reluctant employees to give ideas, how to help employees come up with more and better ideas, how to help employees build on their ideas and develop them further, and how to ferret out the larger implications of seemingly small ideas.

When employees see management responding to their ideas, the organization as a whole becomes much less resistant to change. The ability to listen to small ideas creates a more flexible, responsive, and adaptive company.

The real bottleneck to ideas is not the frontline employees, but the poor reception that ideas normally receive from the organization. Acquiring ideas must become part of the work of every manager, and the organization itself must be aligned to support, rather than resist ideas.

Comments From Employees Using the Idea Process

When asked, "What do you think about the idea process?" The following are unedited responses to this question from employees who are using the idea process:

- The process is good because it involves the employees in making changes;
- It helps the company save money;
- The process is great for making associates think and get involved with making process changes;
- Changes are made to my idea without my knowledge;
- Seems to be approved when there is no cost;
- It's really a great idea and helps things work great;
- I don't think we should have to submit two every month;
- It should be voluntary;
- There is some confusion: I'm not sure who gets or who approves them;
- It's good. It's a way to show that you are aware of the things that can make a job easier;
- It is a good program to help you make the

job easier and give you ideas you would not have thought of before;

- It is a good way to let employees share ideas;

- I feel that it has improved the way things are done and made my work area more efficient for my fellow employees as well. Also it has allowed certain items to be accomplished that may never have been thought of or accepted with this program;

- The program can be a valuable tool. It has improved many processes and made work easier and more productive;

- We need more awareness training so that new employees will be aware of the program;

- It has made some processes better by getting people involved in their jobs and saved the company some money also;

- I think it is a very good program. A lot of ideas have been generated from it;

- It has saved a lot of money and has made many jobs easier and more efficient.

Some of these comments are very positive while others indicate opportunities for improvement. Comments like these can identify areas where training is required, supervisors need to be more involved, or other issues need to be addressed. The employees can tell you what's right and what's not. Managers just need to listen, truly, all you gotta do is ask!

ALL YOU GOTTA DO IS ASK

Chapter XVII

The Principles:

Till

*"People often say that motivation
doesn't last. Well, neither does bathing; that's
why we recommend it daily."—Zig Ziglar*

Promotion of employee involvement is part of every manager's job. Communicate publicly through high profile celebrations and in company newsletters. Have an idea appreciation party and thank employees for their ideas. Communicate on a more personal level during a performance appraisal or through a thank you card or note.

Put all of the ideas implemented during the month into a barrel and pull out a few names for a special prize. At a Dana Corporation plant in California the manager announced in advance that two tickets would be given to the next Raider football game and the plant received the most ideas ever that month.

Managers must keep the enthusiasm going.

111

ALL YOU GOTTA TO DO IS ASK

Following is a list of how some managers and front line supervision describe their role—mostly active, motivational roles:

- *Share;*
- *Contribute;*
- *Facilitate process;*
- *Make process easier;*
- *Implement;*
- *Produce;*
- *Simplify;*
- *Support;*
- *Empower;*
- *Communicate;*
- *Review non-implemented ideas.*

To keep up enthusiasm, keep the employees involved throughout the process. The person who originally suggested an idea should be present to closely follow the results. Not like the old suggestion system where a person submitted an idea for someone else to implement. Rather, employees should submit their ideas and then take leadership to see that they are implemented. Give them ownership.

Make suggestions visible by asking employees to identify and display their problems. Create an opportunity board, which encourages everyone to post up problems and opportunities. Get people involved in identifying and selecting the opportunities.

Listen to complaints. Many are valid problems that can become improvement opportunities.

A newsletter can share and promote the best ideas.

Imagine how people feel when their ideas are included.

Encourage employees to carry a notebook with them. Leonardo da Vinci kept notes throughout his life.

Have a "Wall of Fame" to display photographs that recognize employees who implement ideas, based on frequency of participation.

Sponsor a contest in a certain topic area, such as safety. People can vote on the best idea or have a random drawing among contributors for some prizes.

Enter ideas into a database so you can refer back for further implementations. If possible, the employee or team that created the idea should quantify the benefit or potential benefits.

Comments from Employees Using the Idea Process

When asked, "What does it mean to you?", the following unedited responses from employees who are using the idea process were received:

- *It is very important to the company;*
- *You can learn about other employee's ideas and use them to improve your job;*
- *It means the company wants to hear my ideas;*
- *I am helping my job and the process of doing things;*
- *It means a lot. It gives me a chance to make our jobs easier;*
- *That our company is concerned about how we think we can improve. Our company is*

interested in what we think about our jobs and feels that I can make improvements;

- *I take pride in my workplace;*
- *To help expand your mind in the workplace;*
- *The product is produced faster;*
- *It's a great way to exercise you own thoughts and ideas on how your job can be done better and safer;*
- *It shows that the company values employee's ideas;*
- *A possible way to keep our facility running in the future;*
- *Job security, better work environment and quality;*
- *It can make our jobs easier and save costs.*

Chapter XVIII

The Principles:

Fertilize

"First they ignore you, then they laugh at you, then they fight you, then you win."
—*Mahatma Gandhi*

Implementing improvement ideas should be part of everyone's job. It should be included in job descriptions and be part of the performance review process. Challenge people to think of improvement ideas. Be specific, define problems, suggest ways to address concerns, share employee ideas, encourage, encourage, encourage. The role of a leader is to get results. You will be pleasantly surprised by the ideas your workers have.

Debra Benton in her book, *Executive Charisma: How to Win the Job by Communicating With Confidence*, says, "By asking someone a question, you're adding to their self-esteem and that, in itself, is important in the workplace." She contends that a person could spend a full day at work, do nothing but ask questions, and look good in

115

the eyes of their boss.

A manager can encourage their employees by:

- Asking for ideas;
- Clipping newspaper or magazine articles about how other companies and other industries operate. Pass them around so your people can see them;
- Starting a book club or a discussion group.

Reading a good management book can give you a fresh perspective on things. However, it is a rare person who can read a book and then afterward apply the ideas in their life. Getting together with a group of other people who are reading the same book, may generate questions such as, "How can we apply this information here in our company?" Book clubs can also lead to positive change.

In November 1981, Jack Warne, former CEO of Omark Industries, bought 500 copies of Dr. Shingo's "Green"[6] book and gave them to all of his managers and engineers to read. They split up into small groups, read a few chapters, and asked each other, "How can we apply this new knowledge at Omark?" Within one year, Omark became the best Lean Company in America.

—Look beyond your job—other departments, places you visit. What rule needs to change?—

Back in the 1980s, I led industrial study missions to Japan. On each trip a group of around 20 American managers would visit 12 to 16 different Japanese

---

[6] "The Study of the Toyota Production System from an Engineering Viewpoint." This was the first book in English to move Lean manufacturing forward in the West.

116

companies. For many of the travelers, it was the most important learning experience of their life. They were able to visit world-class companies such as Toyota, Canon, and Sony and ask questions of other managers and employees. Afterwards they would walk the factory to see with their own eyes the actual changes being applied.

We might read a book about the Toyota Production System and still feel that their system doesn't apply to our company. But when we see it right in front of us, it becomes real. We begin to realize that it is not a mystery. We realize that if they can do it, so can we. The people and machines there are very similar to ours, operating at a high level of efficiency. These kinds of benchmarking visits to other companies are vital to recharge our energies. Most of us love competition and when we see what others have done, we feel more assured that we can do it too.

I recently interviewed Gary Convis[7], President of Toyota Motors Manufacturing in Georgetown, Kentucky, and he reiterated how important it is for Toyota managers to visit other companies. In fact, Toyota, on a regular basis, sends its employees out to visit its suppliers. And Toyota encourages its suppliers to send people to spend a week at Toyota to see how their parts fit into the Toyota automobiles.

Another great idea I garnered from that interview is that Toyota employees rotate their jobs every two hours!

Visit other companies to see what you can transfer to yours. Workers are always visiting supermarkets, restaurants, stores, and movie theatres, etc. It is from these

---

[7] Interview is in the book Kaikaku The Power and Magic of Lean by Norman Bodek, PCS Press, Vancouver, WA.

experiences that they learn new ways of doing things.

Managers and supervisors note that persistently thinking of creative ways to generate new ideas have reaped unanticipated positive benefits. They have seen the process:

- *Empower people;*
- *Spark innovation;*
- *Encourage continuous improvement;*
- *Formalize thought process;*
- *Encourage standardization;*
- *Promote better customer service;*
- *Demonstrate more effective organization;*
- *Show teamwork;*
- *Change with the times;*
- *Benefit both the company and the employee;*
- *Improve the organization;*
- *Simplify job tasks.*

An idea management system must become part of the culture. It is the way to harvest ideas and implement improvements, and discard ideas that don't move the business forward.

Comments from Employees Using the Idea Process

When asked, "Has it benefited your company and your fellow employees?", the following unedited responses from employees who are using the idea process were received:

- *There has not been any big injury and production and quality is good;*
- *It has made us a great team;*

- *Some ideas we can use on other lines to help make the transition smooth;*
- *Better quality and production;*
- *Some of the ideas have prevented injuries and created more jobs;*
- *People take a greater interest in their jobs when they see their ideas implemented;*
- *The company has benefited through process improvements that save the company money. It has helped create an environment that is more safety oriented;*
- *It's always good for morale to know that the company cares what the employees have to say;*
- *When good ideas are implemented and not ignored then everyone benefits. Jobs that are labor intensive often become less so, and errors and redundant work processes are often eliminated;*

The company saves time, money and resources. My fellow employees like having some input about their work.

ALL YOU GOTTA DO IS ASK

Chapter XIX

The Principles:

Pollinate

"Of all the things I've done, the most vital is coordinating the people who work with me and aiming their views at a certain goal." —Walt Disney

When managers turn their employees on to submitting and implementing ideas, they **MUST** support and follow up the process. Imagine you see a water pipe burst in the plant or office. You'd want to alert the right people to fix it immediately or take care of it personally before the floor is flooded. How would you feel if nothing was done and your coworkers didn't share your sense of urgency? You might just stop caring in the future. We don't want that to happen. **So look at every idea with a sense of respect, importance, and urgency.**

It is critical that management respond quickly; give feedback to the originator of an idea as soon as possible, maybe by the end of the day; definitely within 24 hours.

121

This will help to reinforce the idea that workers' ideas are important in moving the business forward. Someone should talk to employees about their ideas, thank them, and see how they can be implemented.

Use praise, thank them, show respect, trust, and treat them like adults, the way you would like to be treated.

Years back, I owned a data processing company. I could never praise an employee. I never remember my parents or my teachers ever praising me. They freely criticized me, but never found anything inside me that they felt worthy of praising.

I knew as a manager that I should praise the employees, yet I wanted to be sincere and I could not seem to find the energy to praise anyone.

One day I decided to bite the bullet. I walked nervously over to a data entry operator; took a deep breath, and said, "Alice, you did a great job yesterday and I want you to know how much I appreciate your efforts." As I said those words tears actually came to my eyes. Then I looked into Alice's face, and wow, the love, surprise, and appreciation that came back to me was overwhelming. In the past, I had wanted to feel it first. If I praised, I wanted it to be sincere, but I could never feel it in advance, so I never praised. I didn't realize that I could praise first and the feeling would come afterward. It was a great realization for me.

I understood for the first time in my life what I had been missing all of those years. By not openly praising others, they were the losers, but I was truly the biggest loser.

One of the goals of a manager must be to bring closure to every idea, or it will fester like an open sore. Employees will lose confidence in the process if they don't know what is happening to their ideas. Not getting closure is what kills many suggestion systems. Closure is making sure that every idea submitted is either implemented or the original suggester understands clearly why the idea has not been implemented. Of course, this can be a delicate issue because most people shy away from criticism. If an employee submits an idea and doesn't hear anything more about it, they may become discouraged and may not submit another idea again. If the idea is rejected without a clear understanding of why it was rejected, this can also have a negative effect.

While I believe that all ideas are great, some of them may not be able to be implemented immediately due to money constraints or some other clear reason. Make sure the suggester understands the reason the idea was not implemented. Tell them that the idea might be put into effect at another time or that the idea needs to be developed a little bit more.

Use humor when you explain a rejection. Why do many speeches begin with a joke? Because laughter puts people at ease and makes them more receptive, more responsive, and more attentive. Humor is one of the most effective ways to communicate. I'm not saying that you should make a joke of the feedback; I'm only saying that some well-placed humor to accentuate some key points could prove to be effective.

Similarly, make sure you follow up on suggestions that have been accepted by engaging the workers who came up with the ideas. Have them take leadership in seeing that their ideas are implemented. Even when the idea is beyond the scope of the worker, the suggester should still have the responsibility to see that the idea is applied. This takes away the burden from supervisors and managers, and the worker assumes responsibility rather than unloading it to someone else to implement their idea.

Encourage employees to come up with ideas that directly involve their own work, because the ideas are then easier to implement. Many suggestion systems have been overburdened and then abandoned due to the difficulty involved in trying to implement ideas that involve work in areas of responsibility other than the suggester's. The main intent of bringing the suggestion box to the worker is for the worker to come up with ideas to make their work easier and more interesting. The ideas should relate to the worker's job and to their area of responsibility.

Some employees may suggest good ideas that are outside the scope of the idea system, such as:

- "We need more parking spaces!";
- "The parking lot needs better lighting.";
- "There should be Columbian coffee served in the lunchroom.";
- "I should get a $2 per hour raise.";
- "They should hire a new supervisor.";

They all might be good ideas, but they are really outside the scope of the idea system. You might like to provide some vehicle for employees to express these kinds of ideas, but clearly state the scope of the suggestion system. Tell people that their ideas should come from their own area of responsibility, such as the following:

- *Make their job easier;*

- *Make the work more interesting;*
- *Build you skills;*
- *Reduce costs;*
- *Improve quality;*
- *Improve safety;*
- *Increase throughput—reduce the time it takes to do something;*
- *Improve the environment;*
- *Improve customer service;*
- *Improve your problem solving skills.*

You will find that the workers faced with problems in their own area of responsibility will rise to the challenge. They will find creative solutions and implement their ideas by themselves or with their fellow team members. They will take the leadership to implement their own ideas.

To encourage people to submit ideas, report the results. Place displays in the workplace to inform everyone about implemented improvements and inspiring new ideas. Designate a person as a driver for the process. The driver can remove barriers to get ideas implemented, and make sure results are tracked.

Also take pictures of the problem before and after the change was made. These pictures are powerful sharing devices to help other employees see exactly what was done.

The idea process leads to a positive, high-performance culture. As employees see their ideas being used, they begin to feel more valued as part of the team. They become more involved. They are trusted more as they participate in improvement activities that benefit fellow employees and the company. They feel respected. This all leads to increasingly better ideas. Managers gain greater respect for their employees as they see the quality and quantity of ideas that surface. Employees are trusted with more information, training, and authority.

Handle ideas quickly, effectively, and smoothly. Encourage people to copy and share ideas. Employees like to be involved, so empower them to implement their ideas. Don't stand in the way.

Some managers and supervisors described the following as possible problems that may surface in organizations as they begin the process:

- *No feedback or late feedback;*
- *Poor/Lack of communication with employees;*
- *Poor/Lack of sharing;*
- *Poor/Lack of exploring;*
- *Poor/Lack of understanding/communicating – employees stating, "What's in it for me?";*
- *Lack of recognition, people want praise;*
- *Idea may be too costly;*
- *Idea may be outdated;*
- *Idea may be too complex;*
- *Idea may not be safe;*
- *Idea may not be communicated (properly);*
- *Idea requires another department's involvement;*
- *Coercion – "I need two ideas, right now!";*
- *Speed of implementation;*
- *Number of implemented ideas;*
- *We don't harness people's talent – "Machine down and people wait";*
- *We don't write it down – it's too easy;*
- *Some are just observations (may require*

> *better defining of the process);*
- *Some are ideas for others to do;*
- *"We don't do enough" to encourage idea sharing/copying.*

Comments from Employees Using the Idea Process

When asked, "What obstacles are there to submitting and implementing ideas?", the following are unedited responses to this question from employees who are using the idea process:

- *It is sometimes hard to come up with two ideas per month.;*
- *I see many things throughout the month but fail to write them down and before you know it it's the end of the month and it's a scramble to turn them in. I'm the obstacle;*
- *Depending on your idea you might have to get someone from another area or shift to help;*
- *Follow through is difficult;*
- *No response from top management;*
- *I don't have the authority to implement some of my ideas.*

We want to get, on the average, two ideas per employee each month. The system should be voluntary; some employees may never submit any ideas, while others may give many. In a recent workshop a participant said he submitted 30 last month. Management must respond, offer support, and encourage creative ideas each and every day.

ALL YOU GOTTA DO IS ASK

Chapter XX

The Principles:

Sow & Harvest

"You cannot teach people anything, you can only help them find it within themselves." —Galileo

Idea parties: One plant, on a biweekly basis, stops production for 15 to 30 minutes to enable the employees in the area to brainstorm and write down their ideas.

Stopping People from Working?

Whenever workers on the line at Toyota detect a problem in quality, safety, etc., they are empowered to stop working, to stop the entire line—to stop everyone in that plant from working. When workers detect a problem, they press a button or pull a cord, lights go on, and buzzers and bells go off. Supervisors and other workers immediately run over to help. This is done to get to the root cause of the problem immediately. This system puts enormous pressure on workers to resolve problems. It also gives the workers tremendous trust and respect. It is the essence of the Toyota System—respect for humanity. It is also a great moment to record the problem and solution to share with other workers.

Supervisors must encourage ideas by sharing thoughts from coworkers. The manager's walk, as described earlier, is one way to keep this message in front of people. Even if supervisors don't carry out a structured manager's walk each day, they should keep the idea generation process in front of their employees every day.

The manager's walk can be used to plant different seeds each day, with such topics as:

- *Improve service to customer;*
- *Simplify operations;*
- *Reduce wastes;*
- *Money;*
- *Materials;*
- *Energy;*
- *Time;*
- *Improve procedures;*
- *Better equipment utilization;*
- *Housekeeping;*
- *Safety;*
- *Security;*
- *How to prevent machines from breakdowns;*
- *How to make the job more interesting;*
- *What could cause a potential problem.*

Radical change takes place when managers begin encouraging and implementing large numbers of employee ideas.

Managers need to determine who are giving ideas, who are not, and why. Sometimes entire categories of employees submit relatively few ideas, indicating a deeper, more systemic problem such as communication or cultural sensitivity issues. Some workers can't write English. Ask a bilingual worker to help them write their ideas down.

Direct your training to increase idea generation. Give knowledge in areas where ideas can have the greatest impact. Toyota's training programs help employees come up with many more ideas. They focus on quality, productivity, and safety. Teach employees Poka-yoke,[8] 5S—rigorous housekeeping, lean concepts, quick changeover, Total Productive Maintenance, etc.

A worker on a line in Japan was adding small parts to a sewing machine. To ensure accuracy whenever he pulled a part from the bin in front of him, his hand would touch a small lever, which electronically indicated that the correct part was pulled in the proper sequence to be inserted into the product. What a great idea to prevent defects from occurring. Review ideas for additional potential; even a minor change in one place can create the need for adjustments in other places resulting in more ideas. What else can the idea be used for? Where else can it be used? Who can use the idea? Send it to them. Create a database of replicable ideas. Copying is as good as initiating an idea—get to root causes.

Following is a list of how some managers and front line supervisors further describe their role:

---

[8] Poka-yoke means mistake-proofing, very simple devices discovered by employees to absolutely prevent defects from occurring.

- *Eliminate hurdles;*
- *Motivate;*
- *Train;*
- *Show it can be done;*
- *Influence;*
- *Elevate (to higher management when required);*
- *. Congratulate;*
- *Encourage;*
- *Be a role model;*
- *Reinforce;*
- *Quick feedback;*
- *Show appreciation by saying, "Thank you!"*

Harvest much more than improvements. An idea system liberates people and transforms the way organizations are run. It changes the nature of the relationship between managers and their employees.

As people implement ideas it fosters real respect. Employees can make managers look awfully good. Ideas are the engines of progress. They improve people's lives by creating better ways to do things. An idea begins when a person becomes aware of a problem or opportunity, however small. To the people who come up with the ideas, they are simply common sense. Every employee idea, no matter how small, improves an organization in some way. It is when managers are able to get large numbers of such small ideas that the full power of the idea revolution is unleashed.

Employee's ideas are one of the most significant resources available to us. We fail in the way we practice management if we squander them. Every day, all over the world, millions of working people see problems and opportunities that their managers do not. They watch helplessly as their organizations waste money, disappoint and lose customers, which then results in lowering performance. As a result, employees do not respect or trust management and thus are not fully engaged with their work.

As ideas are applied, everything changes, and changes constantly. Change creates the need for further change, and new problems and opportunities are born. An idea is only a possibility. It must be nurtured, developed, engineered, tinkered with, championed, tested, implemented, and checked. Even then it may prove to be only the start of a further evolutionary process that extends, refines, and combines it with other ideas.

> *It's interesting to watch how people move in the airport. When groups of people come to one of the sets of doors, most follow others through the same door. Sometimes a person will go around the group and through a different door and then others will follow. Fast walkers cut in front of those casually strolling along. Some people look in all the stores while others walk by.*

Once the process begins, the generation of ideas expands, similar to the expanding universe in the Big Bang theory. Taking the suggestion box to the worker is the creation of your company's idea universe. Then, involving them in improvement activities is the expansion of that universe. Albert Einstein said, **"Once you can accept the universe as matter expanding into nothing is something, wearing stripes with plaid comes easy."**

Follow the creative energy from the beginning of the universe to each and every worker. The universe originated from that Big Bang around thirteen billion years ago and everything, absolutely everything, has evolved since then. Look around you and marvel at what was created. Where did it all come from? A mystery, but in truth, everything you see is from and part of that creative process—everything is creative. It has to be that way. Everyone then has the infinite creative potential to grow and expand.

Sunday's Sermon

Mary Manin Morressy of the Living Enrichment Center, on television, referred to a survey in which 50,000 people were interviewed to determine how creative they were. The survey showed that at age 45, only two percent of the people were considered to be highly creative. Age 35, also only two percent. At age 25, it was also only two percent. In fact, at age seven only 10 percent of the children were considered to be highly creative, but at age five, it was 90 percent.

In my workshops I sometimes jest, "When I was in kindergarten, school was fun. I played, painted, made friends, and really had a good time. I also thought in kindergarten that I was really smart, maybe a genius, maybe a potential Einstein. My teacher thought I did clever things. But something happened in first grade that caused a complete reversal to take place. Overnight I went from being a genius to a dunce. I was given tests that showed

134

that I wasn't so smart after all. In spelling bees, I was always the first one to sit down."

Some students had great minds to absorb and remember what they teacher said or what they read in books. They had the ability to repeat back to the teacher what was given to them. I wasn't so fortunate. I could never remember dates in history. I wasn't good at listening and memorizing things but I was very good at daydreaming and creating new things—not totally acceptable to my teachers. But I am grateful that I found a niche to serve others well.

At school we are all taught with the same system to learn at the same pace. But we are all different, and some of us just take additional time to flower and grow. The system is not patient enough to allow that to happen. It wants us all to conform. So 98 percent of students lose their creative potential and we are led to believe that they are just not creative. Even worse, we accept that diagnosis, which is just plain bunk.

All of us, just like seeds, need to be nurtured, watered, and given sunlight to be inspired, challenged, and to be loved. I never remember my parents or my teachers ever encouraging me or praising me to aspire to reach what I could become. They wanted me to conform to their image, not mine. Yes, our school system is designed to make everyone into followers. "What will the neighbors think?" is a very powerful force to conform.

But this can be reversed.

Within a month after bringing the suggestion box to the worker, virtually every person in your company will prove to be highly creative. All you gotta do is ask people:

- *"What do you think?;*
- *How can you make your work easier and more interesting?;*
- *How can we reduce our costs?;*
- *How can we improve our quality?;*
- *How can we improve safety?;*
- *How can we improve customer service?"*

In 1968, George Land used a test from NASA to assess 1,600 children for creativity. He tested the same children as they grew up. The results showed that at five years of age, 98 percent were creative; at 10 years, 30 percent; and at 15 years, only 12 percent were creative. He found that among adults (he tested over a quarter of a million), only 2 percent were creative. Land wrote, "What we have concluded is that non-creative behavior *is learned.*"

This means everyone has creative energy and, hopefully, as you read this book and apply the simple principles, magic will start to happen both at the workplace and at home. It can't be any other way.

> *"Throwing away ideas too soon is like opening a package of flower seeds and then throwing them away because they're not pretty." —Arthur VanGundy*

Comments from Employees Using the Idea Process

When asked, "How can we improve the idea process?", the following unedited responses to this question from employees who are using the idea process were received:

- *Keep on brainstorming and thinking of new ways to make everything easy and more efficient;*
- *Give out certificates of achievement;*
- *Everybody needs to know how important they are;*
- *Maybe we can work on the ideas not implemented yet;*
- *Make an idea leader on each line;*
- *Implementation process needs to be faster;*
- *Have group meetings to get more people involved;*
- *Incentives, morale boosters, show each month how many ideas were submitted and implemented;*
- *Have an idea of the month award based on cost savings and merit;.*
- *An effort should be made to let associates know the value of the ideas process with more than just words;*
- *Supervisory feedback to associates on their submitted ideas could enhance the quality and build enthusiasm for further submissions;*
- *Letting people know that their ideas are appreciated and meaningful for the company;*
- *Reward employees for the implemented ideas.*

ALL YOU GOTTA DO IS ASK

Chapter XXI

Making the Most of Your People

"Each of us is at the center of the universe. So is everyone else."—e.e. cummings

Occasionally, when lecturing about bringing the suggestion box to the worker, managers will say, "Our employees do come up with lots of ideas; they just don't write them down." The problem with not recording ideas is that it is an inconsistent method. It means that management is probably not getting an average of two improvement ideas per worker per month. And these unrecorded ideas are difficult to record and quantify.

Quantifying

Quantifying implemented ideas will add enormous power to the process. In one instance, Claudia Washington, who packs product with bubble wrap, came up with an idea, expanded it, and the process was recorded, and results were quantified. They were tangible. The problem was that the bubble wrap was all over the floor and periodically she or others would trip over it. Bending was difficult, possibly leading to back injury. One day, she had an inspiration. She designed a simple metal fixture to pick up the bubble wrap

139

from the floor. She spoke with her supervisor and a mechanic to help her build the fixture. It worked. She no longer had to bend and was no longer worried about tripping. Great idea!

Then her supervisor, Ken, noted what Claudia had done and suggested replacing the bubble wrap with cheaper wrapping paper. It worked. The same metal fixture was used on the new wrapping paper, which not only helped make Claudia's job easier but the eliminated bubble wrap reduced shelf space by 90 percent. Instead of two semi-trailers coming each week with bubble wrap, only one skid of wrapping paper was required. Claudia and Ken's ideas made their work easier, saved space, and also saved the company over $109,000 a year.

Collect and post the following information on a monthly basis to show the quantified benefit of everyone's ideas:

- Average number of employees;
- Number of ideas submitted;
- Number of ideas implemented;
- Number of employees submitting one idea (and percent of total);
- Percentage of employees involved;
- Number of employees submitting more than one idea (and percent of total);
- Economic benefit from the ideas per employee (need to quantify the ideas);
- Investment per employee in training.

Making Six Sigma more effective

It's interesting how different organizations embrace

improvement efforts. One of the latest efforts is called Six Sigma, created at Motorola in 1986. Six Sigma is a continuous improvement methodology that is data driven, focusing on customer requirements. Companies using Six Sigma certify leaders as master black belts, black belts, etc. These are project leaders who have evolved through a series of levels starting as yellow belts or green belts. Each project follows a series of steps commonly referred to as DMAIC (Define, Measure, Analyze, Improve and Control).

While this system is certainly beneficial, it focuses on the project leaders, not the employees—the idea generators—the strategically important essence of new ideas. Programs like Six Sigma may address some quality issues related to customer requirements but they do not always foster critical problem-solving skills. Training only project leaders could be short sighted if it doesn't create better shared thinking with all employees.

Project leaders and managers may have some of the answers, but not all of them. Managers must instead, of course, look to the employees. They must ask, "How can my workers help others with the same problem?" They must encourage their workers to help each other, like alcoholics helping other alcoholics at Alcoholics Anonymous, or people with weight problems helping others at Weight Watchers.

While managers play a vital role, employees are the experts in their 25 square feet of space. The employees do the job every day and from experience they develop great expertise. Managers must tap into that bank of experience and knowledge.

One great trick to develop employees is to tap into lunchtime chatter. What do your employees complain about over lunch? Sometimes they talk about problems everyone knows about; everyone except for management. Managers would learn a lot if they could listen to employees talking at lunch. Unfortunately the conversation often changes when management is present. Use that creative time to plant a seed. Let them know that you respect their right to have conversations about issues that bother them. Encourage them to talk about and analyze their problems while they are eating lunch. Ask that after their lunchtime discussions, they write down and submit their ideas for improvement.

When employees asked, "What is a valid idea?" the following themes are repeated:

- *Ideas for improvement;*
- *Saves time;*
- *Saves costs;*
- *Improves safety;*
- *Reduces waste;*
- *Increases service to the customer;*
- *Adds value.*

An effective process doesn't concentrate only on problems and how to fix them. The focus should also be on what is being done well, and what can be done to ensure that things are working well more often in more areas.

Employee involvement allows all people to implement improvements, not just those with green and black belts. The process is to empower workers to make their work easier; more interesting; to reduce costs; and improve quality, safety, throughput, and customer service.

Management should develop the hidden talents of

all workers and have all involved in continuous improvement activities. It allows people to wake up in the morning and look forward to going to a work environment where they feel valued and contribute to their work group, the company, the society, and to their families.

"Everyone has an invisible sign hanging from their neck saying, 'Make me feel important.' Never forget this message when working with people." —Mary Kay Ash

ALL YOU GOTTA DO IS ASK

Chapter XXII

Asking Better Questions

An Interview with Rob Curtner

"A question is the midwife to new ideas."
—Socrates

We asked Rob Curtner, an experienced organizational change consultant, to share his ideas on questions in the workplace. Rob is the founder of Learn-2-Learn, a company focused on organizational learning through human interaction.

Yorke: *With your background in both counseling and training and organizational development, your view on what makes for good business communications is focused on people to people interactions and the patterns of these interactions. What are the issues related to asking better questions and why is this important for leadership performance and execution?*

Curtner: You are asking a question about asking questions. Not all questions are created equal. "Hi, how ya doing?" gets the same response every time. Asking better

145

questions gets better information from others about organizational functioning and this can lead to better decisions, and create connections between associates at all levels. By learning to ask better questions, we are taking control of at-work communications at a deeper level.

Bodek: *So true. I remember when people would greet my mother and say, "Dorothy, how are you?" My mother would then tell them about the pains in her side, her lumbago, her gas pains, how uncomfortable she was sitting, that the weather was really affecting her, that she wished the sun would come out, how she reacted last night to the pickles, and on and on. While all the person wanted to hear was "Hi," and then move on.*

Yes, there is a real art to asking questions to further continuous improvement.

Yorke: *What about the use of questions in problem solving?*

Curtner: One definition of problem solving I like is this: "*Problem solving* is asking questions to gather, sort, organize, analyze, and make decisions about non-standard events." Most problem-solving "tools" or forms, where you fill in the blanks, are just a list of questions. The trick is to ask the right questions in the right order. Asking better questions results in better problem solving. Leaders who demonstrate the use of powerful problem- solving questions teach associates to use their methods. This can get the leader outside of the process and empower the associates to do the problem solving on their own.

Answers to questions about "what, where, when, and how much" are important and provide basic facts about the situation.

Bodek: *My opinion differs only slightly. When you have a list and sequence of prepared questions, you lose some of the spontaneity. Often we ask questions to lead people to those answers that we already know. Questioning should be designed to probe the unknown, to bring out the knowledge from both the questioner and the person being asked. Questioning is a learning experience for both. A question should be open-ended, where you do not know the answer and then the next question can help probe even deeper thoughts about the problem.*

Also, the more specific the question , the better, such as, "Harry, did you notice that the glue seems to congeal at the tip of the blade? Do you have any ideas how to solve that problem?"

Or, "Alice, I am sorry that you were burned by the hot pipe. Do you have any ideas what can be done to prevent that from happening again?"

Yorke: *What about the use of the "why" question?*

Curtner: Two-year-olds are learning how to predict what will happen next. They ask "Why?" to learn from adults about what causes certain things to happen. Children often pursue a string of "why" questions in order to take something deeper, to get to the bottom of things. They express a confidence in the parent's knowledge base and often pose some really interesting questions. "Why" is, of course, a powerful question.

Bodek: *Dr. Shigeo Shingo developed this "why" process almost to perfection. He called it the Five Why Method. "Why is the flange not working?" Because the grease has hardened. "Why has the grease hardened?"*

Because the temperature is above the freezing point. "Why did the temperature rise?" Because the air-conditioner stopped. "Why did the air-conditioner stop?" Because someone did not replace the fuse. "Why wasn't the fuse replaced?" Because we didn't look at the checklist. "Why didn't we use the checklist?" We keep doing this until we get to the root cause.

Yorke: *I was first introduced to the Five Whys when I became certified as a Kepner-Tregoe Root Cause Analysis Program Leader. I marveled at its simplicity and found Five Whys to be one of the best tools in my problem-solving toolbox.*

Many times I find people using the Five Whys incorrectly, but even then sometimes it is valuable. Some people use Five Whys as the five excuses, although Five Whys will still lead them toward other possible problems or concerns. I once facilitated a root cause analysis session where the Five Whys were being used as the five excuses. Instead of being a waste of time, this led to the investigation of seven additional problems. After some investigation and the correct use of the Five Whys, seven issues were addressed, all of which when combined and looked at from a distance, appeared to be one problem.

The correct way to use the Five Whys is to answer each "why" with the absolute reason why something occurs, the cause of the problem. Continuing to ask "why" takes the cause to the next level, until eventually the root cause, the real cause of the problem, is found. There is no magic to the number five; sometimes only three "whys" are required, other times six, but usually the root cause can be reached in five or less "whys."

Shigeo Shingo never accepted excuses for problems.

He knew the value of finding the root cause and felt the Five Whys was an effective way to find the true cause of waste. He taught this method to Taiichi Ohno, Vice President of Manufacturing, who helped incorporate this valuable tool into improvement efforts at Toyota.

Supervisors can use the Five Whys to encourage their people to think about the true causes of waste and ask for ideas to reduce or eliminate waste. I once facilitated a group of 12 people in a root cause analysis session. Using the Five Whys we arrived at the root cause, but did not have a way to eliminate it. Our analysis of the problem allowed us to find ways to reduce some of the wastes caused by the problems. Ideas suggested and implemented included a "field fix" (a fix after the product was delivered to the customer), a short-term intervention, and a long-term improvement effort.

Bodek: *There is always a root cause. Often we see a problem and immediately have an explanation and reason why the problem occurred and just accept that problems will always occur. Instead, we must realize that all defects, as an example, can and will be eliminated if we just get to the root cause.*

Never accept "impossible," or "that is just the way things are," as an answer. The Bible says, "Because of your little faith, for truly I tell you, if you have faith the size of a mustard seed, you will say to this mountain 'Move from here to there' and it will move and nothing will be impossible for you."

Yorke: *It seems to me that leaders have a responsibility to lead the questioning process, to inspire others to join in the dialogue and inquiry. How can asking better questions create an organizational climate of*

149

successful information sharing?

Curtner: Leading by demanding information is less effective than leading by thoughtful inquiry. Since all questions are not created equal, some questions have the ability to not only generate useful information, but to develop trust and understanding at the same time. The Organizational Learning approach promotes inquiry as a process of asking questions to work down the Ladder of Inference to uncover the underlying experiences, beliefs, and values that influence a person to think a specific way.

Bodek: *Yes, but the questioning process only works when we are able to overcome people's reluctance to change. So we ask with the confidence that the worker has the answer inside them. And yes, it does take a certain amount of persistence, and faith that with time, the right answers will come out.*

Yorke: *What are leaders really trying to get at when they ask questions?*

Curtner: One feature of questions is the give and take aspect. If I ask you a question, there is an expectation that I will pause and wait for your answer. That means I have to accept the time it takes for you to consider my question, formulate your answer, and deliver your message back to me. If I ask and then do not wait for an answer or if I interrupt your answer, the message conveyed is that what you say is not respected. Why bother to ask at all? There is a simple honesty about asking questions. For all of us, when we are not heard, our trust suffers as well as our motivation. Listening is important, but is often poorly practiced; we are all preoccupied with our own needs. Asking questions assumes responsibility for listening to the answers. This seems obvious.

Bodek: *I was such a poor listener in the past. I would always ask a question knowing the answer already. I never had the patience to wait. I would ask the question and then ten seconds later give the answer. It was like I was having a dialog with myself.*

Yorke: *What could be the bottom-line impact of better questions on a company's performance? What factors influence a company's strategic choice to develop the use of better questions?*

Curtner: Many initiatives dealing with quality, cost savings, lean manufacturing, safety, or preventing sexual harassment are, in a real sense, disciplines requiring the use of specific questions. When an organization begins the new initiative, it is critical for leaders to begin to internalize the correct use of the key questions the initiative requires. Sometimes attitudes may have to shift for the questions to be successfully asked and answered.

Too often the emphasis is on the answer and the action, not on asking the right questions. For example, take the question, "Why do we have so much scrap?" The answer could be, "Because our quality inspection works so well." If we ask, "What are the key causes of our scrap?" the answer may become, "Temporary employees do not receive enough training," or "Lead operators have no disciplinary power." Asking better questions is a major part of any strategic initiative. What questions? Who is asking?

Bodek: *I believe we first have to overcome the stigma attached to questioning. When we ask, "Why did you do that?" or even "Why did it break?" Most people automatically say, "I didn't do it."*

151

Also, there are many questions that are not really looking for answers. "How is the food?" "How was our service?" All that is looked for is, "fine." They don't really want your views; they just wanted to hear that everything is all right. We are then very reluctant to tell waitresses or waiters that the food wasn't as good as expected. We don't want to argue, we don't want to make the day unpleasant, so we just give the answer that the other person expects.

But at work we are looking for continuous improvement, so our questioning process has to overcome the natural predisposition to ready-made answers. We really want to know how to improve and we know that the worker is the real expert on the job and that the worker knows the work better than anyone and that we really want their help to make our company better.

Yorke: *How does a company encourage questioning?*

Curtner: Questions, like other forms of business communication, flow downhill better than uphill. Questions that are critical about the organization's functioning can be heard as complaints, demands, accusations, or cries for help, depending on the direction the communication is flowing.

As I mentioned before, good questions invite thought, requiring more than simple recall. Good questions usually have a variety of correct answers. "Which way did you choose to figure it out?" would elicit a discussion of strategies. You are intrigued to answer a good question; you've never answered this question before, and you don't know exactly what you think about it until you put together your reply. A good question invites and evokes creative, critical thinking. A checklist is a list of questions. Someone has to determine and write down the questions that are critical for others to understand how to ask and answer. The initiative starts with agreement on what questions are needed to ask and teach others to ask.

Bodek: *We are bringing the suggestion box to the worker because we know that important improvement ideas are within all of the workers. To draw out some of the ideas, we can ask simple questions such as, "Tom, our defect rate is stuck. I thought you might have some ideas on what we can do to improve. Have you noticed anything that you think we could try that we might not have thought about before?"*

Yorke: *What infrastructure should a company put into place to help develop better questions? Are there any benefits to using full-time or part-time question expert facilitators? What training should be provided, and what different roles do executives or others in the organization play?*

Curtner: The better use of questions needs a minimum of structure. Job aids, problem-solving tools, process descriptions, quality standards; all can include the key questions that lead to shared high levels of performance. It could be rewarding to ask a number of associates who they think are good at asking questions. Who asks the questions that really matter and cause you to think? Find out whom these people respect and find out what questions are being asked. If no resource like this exists, train supervisors, lead operators, or maintenance people to ask the questions.

Learning to ask good questions takes practice. If you're new to asking a specific line of questions, pause before you ask the second question and wait for an answer.

Here are some guidelines for asking good questions:

- Ask open-ended questions, not yes or no questions. Those will get a one-word response and give you little information. They close off thinking rather than helping others to think into their own problems;
- Don't restrict answers by asking questions that can be answered with a list, unless you want a list;
- Instead, use phrases like: "Tell me all you can." "Describe as completely as you can what your feelings were;."
- Be careful how you use "why?" because it may come across as an accusation, feel like an assault, or put the person on the defensive, like when a parent asks (demands) "Why did you do that?";
- Ask just one question at a time. If you ask two- or more-part questions, you will probably only get the answer to one part;
- Don't interrupt, put words in the person's mouth, or anticipate his/her answers. You may be wrong, and even if you are right, it will be disempowering;
- Use common vocabulary. Don't talk over the person's head, and don't talk down to the person. Know the individual's level of education;
- Don't be too blunt.

Yorke: *What other types of questions are useful?*

Curtner: Let me identify several choices:

1. **Big-picture questions**—Big-picture questions intend to let people step back from what they are doing to understand the negotiation and problem solving within a much larger sphere of activity. There is sometimes a reluctance to ask these questions for fear of the answer, or because the answers themselves are so elusive. And yet, the answers can rally people around meaningful activity and purpose. Here are some examples.

- What are we/they/you trying to accomplish?
- What are the priorities?
- What do we/they/you hope will happen or be avoided?
- What common purpose brings us together?
- Do you think there is a better way to make our work run more smoothly and efficiently?

2. **Questions about problems**—In a typical conflict or problem, there is often an obstacle or a misunderstanding that has assumed a life of its own. One strategy for resolving the problem is to initiate a process for understanding it. I have found that the following questions—if genuinely asked and without an assumed adversarial tone—can offer valuable tools to stimulate conversation that is constructive:

- In your mind, what is the problem that faces us?
- What are the obstacles that get in the way of finding a solution?;
- In what ways do each of us contribute to the problem?

- Do you think the way our work is now constructed is achieving the results we each are hoping for?
- What do we each think and assume about one another?

3. **Out-of-the-box thinking**—The good question can encourage out-of-the-box thinking and thereby open new interest and possibilities for discovering fresh ways to understand and resolve problems. Placing creative options into these questions can stimulate energetic and exciting thinking.

- What would it look like if we . . . ?
- What if we tried . . . ?
- What do you think would be the reaction to . . . ?
- It seems as if we've tried everything. What haven't we tried?
- What other options do you think are open for us?

Bodek: *Questions should be simple and direct to stimulate thoughts and actions. We want to focus on the following:*

- *What can you do to make your job easier? What do you need to make your job easier?*
- *What can you do to make your job more interesting?*
- *How can we reduce our costs?*
- *How can we improve safety? Look around and see if you can find some potential safety problems. Look at the floor, think about the lighting, are objects too sharp, is the area*

> *really clean and spotless, what could be a potential problem, what tires you, what gives you energy?*

- *How can we improve the throughput? Yes, we want to get the job done exactly on time. What do you think gets in the way of that happening?*

- *How can we improve quality? Have you been trained on the seven quality tools?[9] If yes, do you use them? Are you part of a quality team?*

- *How can we improve customer service? Who are your customers? Are they pleased with your work?*

It is not complicated. You just have to be persistent and ask those questions every single day.

Yorke: *What prevents us from asking better questions?*

Curtner: Emotions do. When our emotional involvement increases, our ability to invite further inquiry often decreases. When anxiety over financial and timing outcomes, trust issues, and authority and power enter into the conversation, the strategic aspect of asking questions is short-circuited.

The big question is "How can we ask questions in such a way as to empower and align others, while getting

---

[9] Seven quality tools are: check sheet, pareto chart, flowchart, cause and effect diagram, histogram, scatter diagram, control chart. For an explanation go to
http://deming.eng.clemson.edu/pub/tutorials/qctools/qct.htm#CHKSHEET

the information we need?" Asking better questions begins with an awareness that allows us to hear our own questions, before we ask them. Awareness is aided by a symbol (a reminder) of the process, a guide to help us remember to focus on the process and not the people. The symbol for asking good questions is as close as the last time you asked a good question and found out something really valuable.

Yorke: *Can you describe a central approach to using questions in support of a new corporate initiative?*

Curtner: The first thing to keep in mind is about what to do with the answers to questions. It would be a terrible waste to ask a wonderful question and ignore the answer, and yet it happens. Asking questions we already know the answer to may be fine for detective work, but it does not build trust. Try asking a follow-up question to understand more deeply and to determine if the judgments or assumptions you are considering are valid. Probe with follow-up questions.

Get people at all levels in the organization to begin using questions and following up on where the questions lead.

Bodek: *Right on target! Management can be very reluctant to ask questions of workers when they feel the result will only give them more work to do. But when we bring the suggestion box to the worker with the expectation that the worker will implement the solutions to problems, then management will be much more open to get the employees involved in the problem-solving activities. It is why quality circles and self-directed work teams are so successful. The workers are challenged to tackle problems and then empowered to solve the problems themselves.*

Yorke: *What's to be gained from adding a focus on better questions into the workplace?*

Curtner: Leaders, who are already good at asking questions without seeming aggressive, are known as good listeners and are even sometimes well liked by others. A better style and practice with questions can have benefits to performance in many ways. Better questions lead to better problem solving, better sharing of critical information, and increased levels of trust.

Yorke: *What are some lessons learned?*

Curtner: I have seen some key lessons in two areas. First, the "how" is as important as the "what." Asking questions in a neutral way, with genuine curiosity and a desire to listen and learn will elicit useful information and build trust. Second, if the answers leaders begin to hear are troublesome and cause issues to be raised that ruffle feathers, be prepared to make a commitment to resolve things, to make tough decisions. The job of leaders is to take on these issues and lead the business forward. If asking questions gets things on the table that need attention, but are painful; which is worse, dealing with the discomfort or avoiding the issue?

Yorke: *Rob, thank you for your ideas. It's obvious that you've spent a lot of time focusing on productive workplace interactions.*

ALL YOU GOTTA DO IS ASK

Chapter XXIII

Eliminating Waste

"Do not confuse motion and progress.
A rocking horse keeps moving but does not
make any progress." —Alfred A. Montapert

Lean manufacturing is characterized as the elimination of waste. Waste is any activity that does not add value as seen by the customer. It's difficult for people to see waste. A lean enterprise is focused on eliminating waste. Taiichi Ohno, former Vice President of Production at Toyota, listed seven wastes, and other people may have longer lists. Some waste is easy to see: it may be a defective product, a pile of scrap, waiting for a meeting to start (which should have begun 10 minutes ago), or completing a form that is never used, but gets filed (because a paper record is required).

There are tools that can help to identify waste. Maps such as Value Stream Maps and Process Maps show us the flow of materials, a product, or a service. These maps help indicate where to look for waste. Analyze work and see what is value added and what is non-value added (non-value added is something that adds no value to a product, like waiting, moving material, making defects. Value adding is painting, bending material, cutting, and converting raw material to finished goods.) Reduce or eliminate the things that do not add value. It's important for us to find ways to identify waste because much of it is difficult to see. The processes that are currently used, the processes seen on a day-to-day basis, the normal operating procedure, may in fact be waste.

We have seen waste in many organizations that, to the people in those companies, looks normal. The employees say, "It's the way things are." Some examples of wastes are:

- *Defects;*
- *Machines not functioning properly;*
- *Set-ups;*
- *Waiting, watching;*
- *Transportation;*
- *Not utilizing people's talents;*
- *Inventory;*
- *Excess motion;*
- *Extra processing;*
- *Over production.*

We need to look at our work and processes in different ways, or the things we consider normal will never be considered waste. Many of us resemble the character that Bill Murray played in the movie *Groundhog Day,* where he lived the same day over and over and over again. We go to work, we see what we saw yesterday, do what we did yesterday, and tomorrow we'll do the same thing again. We get trapped into thinking this is the way things are and there's nothing we can do about it. In the movie, Bill Murray's character figured out that he was trapped living the same day over and over. Initially, he tried to kill himself since that was the only solution he could see to get out of it. Eventually he figured out that he needed to grow and learn as the way out.

One solution for an organization to get out of doing the same things over and over, especially if those things shouldn't be done at all, is to tap into employee creativity. Different people have different ideas; the person who does the job has a different perspective than those that don't do the job.

Ohno's seven wastes include waiting, motion, and processing all of which can be called waste in the people system. Poorly trained workers, tools that are difficult to use, parts that don't fit well, and poorly developed procedures result in processing waste. People are working hard, but performance could be better.

Processes designed so that people have to walk for parts and tools create motion waste; even reaching for them is a waste. Waiting waste may be the worst of all. In assembly, it means people waiting for the parts they need to work with. Employees waiting for a meeting to start because their leader is busy with other matters result in a tremendous amount of waste. Unnecessary meetings, insufficient resources, unclear work instructions, indecision, unclear authority, withholding information, and improper processes all contribute to waiting waste. Standing and watching a machine is a terrible waste of the human mind. People are our most important asset and yet we allow our employees to wait for meetings to start, sometimes unnecessary meetings. Typically we don't think of this as waste, but people waiting means they aren't adding value.

In the United States it been estimated that up to 97

percent of all our output by weight becomes waste. The accuracy of this figure is not important. If you weigh all your purchases over a period of time and compare it to the weight of what you throw out, donate, and sell, you will find a very real picture of waste.

The point is that we are used to living with waste. This makes it difficult to identify waste. While some wastes, like excess inventory filling a warehouse may be easy to identify, others such as behavioral wastes are not. The lists on the following pages can help identify waste in your organization. Once you see waste, you can work to reduce or eliminate it. Waste includes:

- *Absenteeism;*
- *Accidents;*
- *Administrative tasks;*
- *Arguments;*
- *Attempting too much;*
- *Avoiding decisions;*
- *Being late to meetings;*
- *Being overwhelmed;*
- *Bias or prejudice;*
- *Breakdowns;*
- *Burnout;*
- *Can't say "no";*
- *Changing priorities;*
- *Clutter;*
- *Complaining attitude;*
- *Conflicting priorities;*
- *Crises resulting from lack of planning;*
- *Crowded workspace;*
- *Defective software;*
- *Delegation to wrong person;*
- *Disorganization;*

- *Duplication of effort;*
- *Emergencies;*
- *Equipment failure;*
- *Excuses;*
- *Failure to delegate;*
- *Failure to learn lessons;*
- *Failure to listen;*
- *Failure to voice concerns;*
- *Fatigue;*
- *Firefighting;*
- *Gossip;*
- *Haste;*
- *High turnover;*
- *Idle employees;*
- *Ignoring deadlines;*
- *Improper systems;*
- *Inability to act;*
- *Inconsistency;*
- *Indecision;*
- *Ineffective meetings;*
- *Ineffective tools;*
- *Inefficiency of others;*
- *Inefficient equipment;*
- *Insufficient job training;*
- *Insufficient phone systems;*
- *Interpersonal conflict;*
- *Interruptions;*
- *Junk e-mail and mail;*
- *Lack of authority;*
- *Lack of deadlines;*
- *Lack of decision making;*
- *Lack of empathy;*
- *Lack of focus;*
- *Lack of internal support;*

- *Lack of patience;*
- *Late approvals;*
- *Looking for things;*
- *Low morale;*
- *Meetings that start late or end late;*
- *Messy environment;*
- *Mistakes;*
- *Need to "do it myself";*
- *Negative thinking;*
- *No-shows;*
- *Not enough resources;*
- *Not enough time;*
- *Not respecting other's time;*
- *Office "drop-ins";*
- *Outside commitments;*
- *Over supervision of others;*
- *Over analysis;*
- *Over planning;*
- *Paper shuffling;*
- *Personal disorganization;*
- *Politics;*
- *Poor attitude;*
- *Poor communication skills;*
- *Poor decision making;*
- *Poor delegation;*
- *Poor filing systems;*
- *Poor information management;*
- *Poor listening;*
- *Poor planning;*
- *Poor skills;*
- *Poor team player;*
- *Poor time management;*
- *Poor training;*
- *Postponed decisions;*

- *Procrastination;*
- *Reading unnecessary material;*
- *Red tape;*
- *Redoing something;*
- *Repeated handling of correspondence;*
- *Resistance to learning new skills;*
- *Revised deadlines;*
- *Sarcasm;*
- *Scattered resources;*
- *Sickness;*
- *Slow reading;*
- *Socializing (in most cases);*
- *Spreading yourself too thin;*
- *Staffing problems;*
- *Stress;*
- *Telephone interruptions;*
- *Threats;*
- *Tolerance of poor performance;*
- *Too many meetings;*
- *Too much work;*
- *Travel time;*
- *Turf battles;*
- *Unclear authority;*
- *Unclear job duties;*
- *Unclear purpose;*
- *Unclear/unnecessary memos;*
- *Unmanaged conflict;*
- *Unnecessary meetings;*
- *Unplanned work;*
- *Unrealistic time estimates;*
- *Unscreened phone calls;*
- *Untrained staff;*
- *Withholding information.*

Ohno also identifies defects, overproduction, transportation, and inventory as wastes. Methods are available that address each of these wastes. Anyone familiar with Lean Manufacturing knows these methods and tools, including Kanban[10], SMED[11], Workplace Organization, Standardized Work, TPM[12], etc.[13] The walking suggestion box brings together the people and the processes.

At a Lean Manufacturing conference, Lloyd Trotter, President and CEO, GE Industrial Systems, made an insightful comment, "It's all about an order coming in the front door and a quality product or service going out the back door. Treat EVERYTHING in the middle as waste." This statement suggests that continuous improvement is always necessary.

Rather than looking at the things we do every day as static, the way things are, we need to always focus on looking at those things and questioning whether they are being accomplished in the best way.

---

[10] Kanban - A signaling device that gives instruction for production or conveyance of items in a pull system.
[11] SMED – Single-Minute-Exchange of Die. A system developed by Dr. Shingo.
[12] TPM – Total Productive Maintenance
[13] For more detail on Lean methods and tools please read Kaikaku The Power and Magic of Lean

Chapter XXIV

Sustaining

"You've always had the power"
—the Wizard of Oz

Actively managing the creative energy in the workforce means understanding how it's harnessed and how it's focused. As we mentioned earlier:

1. Harnessing the creative energy in our workers requires a system to tap into the source (people) and capture the creative energy (ideas). In an organization this means having a system in place for this to occur;

2. Someone needs to initiate or introduce the system into the organization. This may be someone at a high level, such as, CEO, President, a managing partner, or Human Resource Manager. The champion or initiating sponsor must have the power to sanction the new system and allocate resources (if required);

3. On an ongoing basis, other champions or sponsors are needed to manage the creative energy. These are the "walking suggestion boxes," which encourage and challenge people. Their tools are their questions and sharing of information. The supervisor must teach the employee how to implement ideas. It is the "walking suggestion box" that becomes the creativity management expert, harnessing ideas, managing consequences, and providing educational opportunities.

It's critical that all people at every level in the organization know that their ideas are welcome. All ideas, even bad ideas, are welcome. We welcome bad ideas because they can trigger good ideas. Every great genius makes thousands of mistakes before the brilliant ideas emerge.

It's tough for people to offer ideas, especially good ideas, if they are not aware of problems or challenges. Supervisors must inform their workers what's important. They must endeavor to make employee creativity part of the corporate culture and actively solicit ideas.

Keep idea generation in front of people. Even if a supervisor doesn't have a structured "manager's walk," he/she must keep the process visible every day. Supervisors should talk with the workers in their work areas, not in offices or conference rooms. They should discuss company and departmental problems as well as goal and objectives. They must encourage ideas by sharing thoughts from co-workers. Supervisors must ensure that ideas are implemented, shared with other areas, and encourage workers to build on existing ideas for continuous improvement.

The secret of good supervision is to empower people. At the airport at Oahu, Hawaii I gave the Northwest Airlines desk attendant my ticket to Japan. She noticed that the portion of the ticket required was missing and the remaining portion of the ticket was from an earlier segment. I suggested that the attendant in Portland, Oregon must have pulled the wrong part of the ticket. At first, the desk attendant stated that regulations required her to charge me $100 for the missing stub, but she would check with her supervisor. A few minutes later she returned and gave me the boarding pass to Japan without any additional charge. She said, "My supervisor empowered me to do what I thought was right."

Thomas Edison, it is said, tried 8,000 to 10,000 ideas for the light bulb filament before he found one that gave him the results he wanted. Not every idea will be successful. Not every idea will be implemented. We need to create an environment where workers feel safe suggesting ideas, either good or bad. To get the results we seek, we need a lot of ideas so we can find and sort out the best ones.

Challenge people to think of improvement ideas. Be specific, define problems, suggest ways to address concerns, share employee ideas, and encourage; encourage, encourage. The role of a leader is to get results; you may be surprised by the ideas your workers have. When people are involved in implementing their own ideas, they find a way to do them quickly. Not only will you get results, you'll get them sooner.

Designating a person as a driver for the process is an effective way to ensure the process continues. Besides providing guidance to managers and supervisors, the driver assists in the promotion of the process. This person ensures that all training regarding the continuous process is held, makes suggestion forms accessible, and assists employees in writing down their ideas. The driver makes the process visible. The driver removes barriers when an idea is implemented, and makes sure results are tracked.

It is management's job to create an innovation process. Leaders must define innovation as a part of the company vision and communicate its importance to the organization. It is management's job to create a continuous improvement process to capture and implement ideas. Management must ensure that involvement is not voluntary, but it becomes part of each employee's job. When first implementing such a process in an organization, it will take some time for supervisors to become engaged. Be patient and don't force a timeline for compliance. Positioning the process within the organization is also a factor in how it will be sustained. Peter M. Senge in his book *The Fifth Discipline* writes, "At its simplest level, a shared vision is the answer to the question, What do we want to create?" According to Senge, a shared vision connects people and binds them with a common purpose.

It is essential for each employee to come to realize that documented ideas for continuous improvement is part of everyone's day to day job. All employees must look at their jobs differently. They must embrace change. They must realize that while 85 or 90 percent of their job is to do the work assigned, the remainder is to find ways to improve their work.

By positioning continuous improvement as part of an organization's day-to-day operation, employees expect to participate, they are responsible for generating ideas and implementing them, and they find it easier to change. All employees realize they have an impact on the bottom line.

When we asked hundreds of employees, "What's in it for you?" they gave answers that were very similar to the goals mentioned earlier by supervisors and others:

- *Makes my job easier;*
- *Job security;*
- *Company can be more competitive;*
- *Better environment;*
- *More control over job;*
- *Get work done faster;*
- *Self-actualization;*
- *Better idea of the way the company works;*
- *Recognition;*
- *Feel like I'm making a difference;*
- *Safer work environment;*
- *I may be able to prove myself capable of a better position.*

Donald Peterson, President and CEO of Ford Motor Company, as quoted by Bob Nelson, said, "When I started visiting the plants and meeting with employees, what was reassuring was the tremendous, positive energy in our conversations. One man said he'd been with Ford for 25 years and hated every minute of it . . . until he was asked for his opinion. He said that transformed his job."

People are often relieved to be in an environment where they feel comfortable making a contribution. There is a story that when Buddha was in a grove, people walking up to the grove lost all their cares and worries as they moved closer. Part of their relief came from the energy manifesting through the Buddha, and part was the very nature of the grove itself. In other words, their comfort was directly related to Buddha's thoughts and the environment. In a work setting, this can translate to the caring interactions of one's supervisor and an environment that encourages participation.

Supervisors need to realize that this is a trust building system, a way to communicate with employees where they work. Share thoughts, problems, and concerns with workers. Seek input, encourage ideas, and assist with implementing ideas.

Chapter XXV

Barriers

*"It is easier to get forgiveness than permission" - **Stuart's Law of Retroaction***

Rhonda Schneider, a director managing various departments, asked her employees to identify barriers to implementing two ideas per month. Their answers follow in italics, ideas to reduce or eliminate the barrier are suggested after each answer:

1. *Everyone does not participate in the program. Some people are shy, or have difficulties in language and in expressing themselves.* The supervisor should ask them questions and help them to participate, but some people might not ever offer an improvement idea;

2. *Misconception about what an improvement idea is.* They still do not know that it is what they can do themselves and not by others. When you show them examples from others they will slowly be able to better participate. Also having your teams talk about the idea process will help;

3. *Lack of follow up until the idea is implemented.*

The supervisors and managers will begin to realize that the idea process is the best and most important part of their job. Isn't it management's responsibility to get "work done through other people?" And the best way to do this is from the worker's own ideas;

4. *Find it hard to submit two ideas per month.* List the names of the top producers and how many ideas they have submitted to encourage others;

5. *"That is not my job!" attitude. How to address this?* The improvement process is part of your job—we need your brawn and your brains—who said you leave your brains at home?;

6. *Why is the improvement idea process mandatory? The associates feel pressured to submit ideas.* Yes, you do want improvement ideas and some pressure (not negatively) is not bad. Some months you won't get any ideas from them. Redesign their work, rotate them;

7. *Our employee handbook does not mention anything about submitting two improvement ideas per month.* Add it to your employee handbook with an explanation. If the company is to grow and be competitive, it needs everyone's help;

8. *The associates are reminded about the improvement idea process only at the end of the month.* Change this. Do it daily! Are your supervisors and leads trained to lead and manage the process?;

9. *Who should be responsible to see that all the associates submit two ideas per month?* Of course, all supervisors and managers are responsible;

10. *They cannot come up with two ideas because of communication issues, especially on second shift.* Maybe just better training is required for the supervisors on the second shift;

11. *No encouragement for the associates to submit ideas outside their work areas.* This is generally true for an individual unless the manager needs wider help;

12. *Need to get feedback on the ideas that were submitted, even if an idea was not implemented.* Of course, daily feedback is vital. If an associate comes up with an idea and it is neglected, then they may become discouraged and not offer any more ideas;

13. *It should be implemented; otherwise what is the point in submitting an idea?* Our associates are discouraged when they submit a few ideas and even those are not implemented. This must be addressed through your attention and questions. Not one single idea is to be ignored. The idea process is your most powerful vehicle. It is the heart of lean;

14. *Should not make the associates feel bad when they do not submit any ideas.* Yes, this is important. People will help; they have the intelligence; they just need guidance and some

177

personal attention. Question them; ask for their help on specific problems; how to improve the safety; how to make this a better more cheerful place to work, etc.;

15. *What is the motivation to submit more improvement ideas?* Need some kind of incentives for submitting ideas. The first reward is that it is part of their job and yes do dream up rewards that you can give;

16. *An idea is marked as implemented but it is not followed through.* Yes, people know when you are playing a game with them. Just to stamp it implemented will be a discouragement. (Every idea is important and valued.);

17. *Some people think it is not important. If they work their 7.5 hrs., it should be sufficient.* You want to change this attitude—you want to be the best company to work at, you can do it from "bottom-up management;" you can do it from their ideas and ideas are unlimited. What is an idea? Remember the smaller the better;

18. *Ideas are not implemented if there are changes expected in the future.* Do it! Then you change again later;

19. *Do not get support from other departments to implement the ideas.* You need better team work, and better buy in when an idea is outside of your area;

20. *The ideas submitted by the office personnel mainly require the IS department to provide*

support, this takes a lot of time to implement. Encourage people to learn how to program the computer. Come up with ideas that you can do. Is your office spotless? Is your work exciting? Are you learning and growing?;

21. *If first shift implements an idea it is not being followed on second shift.* All you need is better communication between shifts;

22. *Individual ideas are being counted as group ideas for the whole team.* This also discourages associates if they have worked very hard to implement it. Their efforts are not being recognized. Let them talk about this until there is agreement. A group idea is good if people really participate in it. You do not want to discourage them;

23. *We need to have an idea committee for the department to see if the ideas can be implemented or not, and if it cannot be implemented they would have to respond to the respective associate about their idea.* Set up idea committees — give people more responsibility. Let people walk through their part of the plant looking for things to improve. Set up a problem board to share needs;

24. *Some of the ideas are implemented for a month and then it falls apart.* This is management. Ask people to come up with ideas on how to solve this.

Once again, some of these comments are very positive, while others indicate opportunities for improvement. Your people will tell you what's right and what's not. Truly, all you gotta do is ask.

Chapter XXVI

Begin The Journey

"A thousand mile journey begins with one step."—Lao Tzu

Remember the people in the airport? Some went into the coffee shop, others into the bookstore or newsstand. Some were casually strolling, while others hustled to the gate or ticket counter. It seemed that each person had a destination or an objective, such as catching a plane or finding something to eat. Each of these people began their journey, not in the airport, but somewhere else, maybe earlier in the day or maybe a long time before.

Woody Allen once remarked, "I'm astounded by people who want to 'know' the Universe when it's hard enough to find your way around Chinatown." I hope we provided a little information to help you find your way in your work and not so much that you'd need to sort it out and decide what to do and how to get there.

Cultural anthropologist, Gregory Bateson, tells a story about a man who asked an advanced computer, "Do you compute that you will ever be able to think like a human being?" The computer processed this question for a while and finally said, "That reminds me of a story . . ."

This story originated as a response to the question of whether artificial intelligence was possible. The response is an interesting observation that people are storytellers. For thousands of years, storytelling was the primary means of teaching others. People effectively communicate through stories. It is one of the ways human beings differ from other living things

One of the things we like about the process in the book, _The Idea Generator – Quick and Easy Kaizen,_ is how the ideas are communicated as a story. The book lays out simple Quick & Easy Kaizen forms, which briefly explain before and after improvement along with the effect. The forms tell a story. When the forms are displayed, others can see the story, which reads: We had this kind of problem or issue. We took this action. We improved a little bit.

Stories are one of the best ways to share information. For management, it can be a way to plant seeds for improvement or a safe way to collect ideas from workers. Michael Abrasoff, in his book, _It's Your Ship!_ tells how as Commander of the USS Benfold, he asked each and every crewmember to share stories of what was right and what was wrong. He also asked for ideas on how to improve. Working with his crew, the USS Benfold became the best ship in the Navy. What Commander Abrasoff did is what every CEO, every manager, every supervisor, and every team leader should do: talk to their people. Not once when starting their "command;" not once or twice a year during performance reviews; but every day.

Usually employees are happy when they get the chance to become more involved, offer ideas, and be heard. Management can put a damper on employee excitement by not handling changes or a lack of commitment to improvement efforts.

A supportive environment is essential for employee involvement. Listen to your people; ask for their opinion and for their ideas. It is important that people feel they have the freedom to propose ideas. Take the suggestion box to the employees and capture new ideas. Convince them that change is possible. Invest the time to talk. New ideas will grow from these conversations. An encouraging culture instills creativity and passion. Employee involvement requires management to listen. It also means those involved will develop the ability to think and do their job more effectively.

A Great Idea

Michael Miller worked on the assembly line packaging product for stores. His job was to close 8,000 to 10,000 case covers a day. Imagine doing Michael's job every day of the week.

When Michael was introduced to Quick and Easy Kaizen, empowering him to make his work easier and more interesting, the "light bulb" went off in his head. He got two pieces of wood, tied them together with pieces of cardboard and adhesive tape, and placed them against the assembly line. This simple contraption automatically closed the case covers.

Michael's idea totally replaced himself. How many

of you would do this if you had the chance?

Recently, in Cleveland, Ohio I shared Michael's idea and asked the audience, "Raise your hand if you would do the same thing as Michael." Out of 50 attendees, only one person raised a hand. Another attendee laughed and said, "Yea, he raised his hand because he owns the business."

Richard E. Dauch bought General Motor's Detroit gear and axle operations and formed American Axle, which has grown to 20 plants from five. The business is consistently profitable. Dauch claims that all he did was change the way people work together.

If American Axle employees had worked the same way as they had for GM, the way it had always been done, they wouldn't have grown. Most likely, American Axle would no longer be in business.

Walking suggestion boxes create a secure environment; workers must feel safe making suggestions and trying out new ideas. In March 2002, at a conference of experts on the Toyota Production System, Nampachi Hayashi, Toyota Executive Advisory Engineer, emphasized that a Kaizen mindset was required throughout the company. This is the mindset required in any company that wants to succeed. Continuous improvement is too important to be left to process engineers or senior management. It must be a part of what every worker does every day.

Power and fear may be the root of traditionally poor idea generation in organizations. As noted earlier, supervisors feel that ideas are their responsibility. "Workers should just do their job and leave their brains at the front door!" Supervisors may be afraid that some workers are smarter than they are, so they suppress ideas. On the other hand, workers are afraid to come up with ideas because they might sound stupid or silly. People are afraid to make mistakes, so they just keep doing what they have been doing because it is safe. And yet, the only way we really learn is through making mistakes and then correcting them, or by trying something new and learning from our success.

To be successful in this highly competitive world, we have to take advantage of all of the talent at hand. If we continue to suppress latent talent, then we can continue to watch the jobs going elsewhere. It's time to stop it!

Let's spend another minute to look at Toyota. This is a company that believes its business growth is a result of the growth of its people. Toyota realizes the value of its people and helps each person improve his or her ability to do the job more effectively. Toyota believes that the person who does the work is the best person to suggest ways to improve the work and solve problems. Each worker takes responsibility for implementing counter measures to problems. And some workers make suggestions that eliminate their positions.

Toyota has created an environment where people feel safe, even when eliminating their own position. Management and workers must trust each other for this kind of involvement to exist. This kind of mutual trust doesn't happen by accident nor is it an automatic byproduct of implementing the Toyota Production System or Lean Manufacturing. This is trust that must be earned. Commander Abrasoff talked with his crew. The walking suggestion box talks with employees.

The process of communicating improvement ideas

185

as a story is the human side of lean. The simple story actually transfers lean ownership to workers at the appropriate process level, where the work is performed.

Remember Claudia Washington, who packed the finished products into the large shipping boxes? She had to bend over to grab the bubble wrap. She had to twist it and rip it. She wasted a lot of motion and she could have easily tripped over the bubble wrap.

Claudia came up with a simple but powerful idea to put together two pieces of pipe to create a stand where the bubble wrap was placed. She no longer had to bend down. It was safer and easier on her back and she could tear the sheets more consistently, saving bubble wrap.

It was a wonderful idea. Imagine how Claudia felt when she came up with the idea to make her own work easier and more interesting. It was a win-win situation for the company and for her.

It may be appropriate to take a quick look at some of the differences in the culture at Japanese versus western companies and then explore the role of management and worker in the continuous improvement process. Cost cutting and the laying off of large numbers of workers typically characterize change efforts in many western organizations. In Japanese companies workers are seen as the most valuable asset. Rather than lay off people, an investment is made to retrain these workers. This difference takes much of the fear out of the workplace. Employees can concentrate on improvement efforts instead of worrying if they'll have a job tomorrow.

Of course, not all western companies (or their leaders) look at the workforce strictly as a cost. Larry

Bossidy, Chairman of Honeywell International, and Ram Charan in their book, _Execution,_ say there are three core processes that allow a company to get things done. These are the people process, the strategy process, and the operations process. The people process is the one that makes sure people are properly developed and able to handle the strategy and the operation's process.

The statement is so true, but rarely done in industry. People's inherent talents are usually wasted. As people are able to express their creative talent, they have greater respect for themselves and for their superiors, the ones that have allowed them to implement their ideas. And management, now seeing the talents emerge from the workers, have so much more respect for them. How can you not respect someone like Michael or Claudia? Reducing the fear of job loss results in deep loyalty from employees. Masao Toda of Suzuki Motor Corporation said, **"Appeal by senior managers to improve quality is more easily understood and accepted by the workforce than calls to cut costs or improve productivity."**

The bottom line is that most western companies view workers as a cost, while most Japanese companies view workers as an asset. Interestingly, people are the only asset that appreciates rather than depreciates.

Recently while talking with a major corporation's human resources representative. He noticed that a lot of newly hired employees were waiting for an orientation meeting to begin that was scheduled to start half an hour earlier. The HR person said, "That's OK, I'm paying them. What do they care?" Instead of calling employees human resources, we should call them human beings. Maybe we'd treat them better!

ALL YOU GOTTA DO IS ASK

Following the principles outlined in this book for employee involvement will enable a company to use its most valuable asset, its people, for competitive advantage. Management, of course, must provide the leadership that allows this to happen.

Implementing improvement efforts on the spot by taking the suggestion box to the worker, at the worker's location, enables the use of the fewest resources possible to make the improvement. This helps promote employee responsibility for the work and accountability. Rapid decision making reduces the waste of waiting.

The goal for both management and the workers is to drive the continuous improvement of the operations from the bottom-up. In many organizations, however, continuous improvement is driven from the top-down, which is counterproductive. Strategy should come from the top, while improvement needs to come from those who do the work.

Watching the people in the airport, I realized that there's no way to control all this activity. People going in and out of stores, some rushing to their planes, others slowly stroll to kill some time. We are all individuals and we all do things in our own way. We can organize this behavior, but we can't control it. We can, however, manage it.

Similarly, we can't control the creativity within the workforce. But we can manage it. Creativity is very unpredictable. However, it can be encouraged and promoted. We are looking for ideas to improve our work. When asked how he comes up with a good idea, Linus Pauling, the scientist who discovered vitamin C, said, "The best way to have a good idea is to have lots of ideas."

Managing creativity is all about creating an environment where we are more likely to be creative. Sharing knowledge throughout the organization helps to create this environment. If an idea is implemented in one area, that's great. If the idea can be shared and implemented in other areas, that's even better. Following the principles we've spelled out allows us to manage this creative process.

Shelly Blakita, a department manager, wrote the following story about creating the environment:

When we first implemented our idea process here, I thought great, another flavor of the month program. This too would pass. However, this program was not a flavor of the month initiative; it became embedded in our management's performance reviews and in signs, classes, and literature around the company.

At first the program encompassed everything. Just submit ideas that you thought would increase production or save costs. I was submitting elaborate ideas that required programming changes or creation of new programs. My other rationale was the idea must be able to save the company a lot of money. So, I was always looking for the big ideas that would save thousands of dollars. This went on for a few months until I went to a seminar called "The Idea Generator - Quick and Easy Kaizen."

The seminar focused on generating ideas that each individual could implement. The ideas could be found

189

within the 25 square feet of space that each associate works in. These small ideas add up to process improvements and most of all in changing the culture to Lean Thinking, continual improvement. As I sat in the seminar and listened, the light bulb went on. "Yes I can do this and we can take this to the floor and really implement it."

I went back to work and not only did I change the way I looked at ideas, but I started to talk with my associates about the program and some of them understood and started to participate. But the participation level was low and it was our busy season. Although I continued to contribute, my focus went to meeting customer deadlines and less to improvement ideas. When our work slowed down, our department began to focus on employee ideas. We bought boards to post the ideas and now we just had to start receiving them.

I started to talk about employee ideas every day in our shift meeting. Every time someone submitted an idea, I brought it up in the meeting. I read the written idea forms, indicated the author, and the impact that implementing that idea had on their work. I would then tack it on our idea board, while the other associates applauded the author of the idea. I would also encourage all the associates to spend some time reading the posted ideas. If the implementation needed outside help, computer support, maintenance, etc, I gave a progress report. Once these ideas were implemented, I would bring them up again and indicated they had been implemented and thanked the person for contributing. I also read my ideas to the associates so that they knew that I was participating in the program. The participation for my shift went from two to four ideas a month to 20. This was all from just talking about it and acknowledging the ideas that were submitted.

In my department we have also had some success with generating ideas by doing the following:

1. *Performing Gemba walks, where we pick a topic and take about a half hour or an hour and walk about and see how we can make improvements for that subject matter. We talk to the people on the lines. When they have an idea we hand them a form or we fill it out and they are given credit for the improvement. These walks have created 10-20 improvement ideas a walk;*

2. *We read some ideas in the departmental meetings and recognize those people for their ideas and their contribution in continual improvement of the department;*

3. *We have also given out recognition awards for individuals who contribute "above and beyond" expectations;*

4. *We purchased more corkboards to post more Ideas;*

5. *We keep idea forms in the conference room. Whenever we have meetings and someone comes up with an idea, that person can fill out the form right away;*

6. *We put forms in holders at the supervisor's area, so associates have "Quick and Easy" access to the forms;*

7. *We also keep the unimplemented ideas and reflect on them once a month. I have been able to pull out quite a few that we can now implement.*

ALL YOU GOTTA DO IS ASK

I asked Gayle Shaughnessy, one of my supervisors, about her success in creating involvement among her associates:

"At first, I wasn't getting any ideas from 99 percent of my associates. I had two associates very engaged in the program. I decided to go to each line and observe. I was sure there were ideas being implemented. Just by observing I began pointing out little things associates were doing to help their job and handed them a form or, if necessary, assisted them in filling it out. It took a few months of pointing out their ideas until the culture began to change and they began writing down the things that they were automatically doing. Now they recognize when they have an idea, they write it down, or bring it to my attention and I or another associate will help them write it down.

"The hardest thing to get across to the people was that they could copy an idea from another associate if they too, implemented it. Now every person who works for me has had some implemented ideas. Another thing that helped this change was taking pictures of improvement ideas and sharing them with other people. By doing this and making it a special event the associates take pride in the ideas they have and sharing them with others.

"One idea in particular illustrates this point. We had a lot of work orders calling for two products to be placed in separate boxes, and then these boxes would be placed inside a long box. The individual boxes were difficult to slide into the long box, because of the flap at the bottom of the long box. One associate took a stiff piece of thin cardboard and made, for lack of a better term, a shoehorn that allowed the boxes to slide into the long box with ease. This simple idea increased our productivity on these orders and the whole line was using them."

Gayle also keeps an idea matrix of all of her associates. This has had several benefits. She will read each and every idea, enter it into the matrix, and if it is something that can be shared, she will forward it to the other supervisors in the department. Keeping track of who has participated allows her to go out and work with those who have been reluctant to contribute. This allows her to work one on one with her associates and build a working relationship with them.

I also spoke with Brian Starkey, an associate who in the last three months has been contributing at least four ideas a month, if not more. Brian said about his involvement, "I participate because it inspires me to think about how I can make my job easier. It also allows me to talk to my coworkers and share ideas with them, builds teamwork. I take my job seriously and it motivates me to be involved with my work. I have implemented several ideas that have saved me time in a process, or evolved around potential safety hazards."

Dale Johnson, another supervisor, indicates she likes our idea program because, "It empowers associates and makes them feel like they are a part of the decision making process." She also says, "There is a synergy that is created by sharing ideas. One person has an idea, another person reads it and adds to it, creating yet a new idea and the process continues. I have seen one idea lead to five or six other ideas."

Really there is no mystery to the process. It is all in how you approach it. If you observe and encourage involvement every day people will participate. If you are excited about the ideas that are being brought to you verbally or written, or if you can help make someone's idea come to life, they feel empowered. It is the people who do

193

the jobs day in and day out that should be participating in the continual improvement of department and company.

If there is a hard part to the whole process, it is sustaining the level of involvement. Continuing to observe and encourage day in and day out, whether it is slow or busy, must be each and every one of our objectives. We are responsible for the success of the company.

Tips to keeping the ideas coming:

1. *Set a goal for each individual, department and facility;*

2. *Make it a part of management performance reviews. It is everyone's job to improve the process;*

3. *Listen to people when they talk. Lots of ideas can be found in listening to issues or solving problems;*

4. *Encourage creativity. Step outside of those paradigms. Just because we always have done it this way doesn't mean there isn't a better way;*

5. *Take pictures and post the ideas where all the associates can see;*

6. *Talk about ideas in meetings, read the ideas and give recognition;*

7. *Talk to everyone who submits an idea about the idea;*

8. *Share the ideas with everyone. If it changes a*

process for the better, change the work instructions and make it standard procedure.

Amazingly, following our principles also further encourages employee involvement while raising the self-esteem of the worker. Implementing one's own idea is a reward in itself. The worker is driven primarily by intrinsic motivation which is much more powerful then an extrinsic motivator such as money.

Today is the day to begin this journey.

If employee suggestions are a good idea and employees won't come to the suggestion box, then we need to take the suggestion box to the employees. Ask questions where the work is done. Reward managers who are not afraid of letting subordinates shine.

It may take a little while for creativity to take hold in your workplace. Start to cultivate the culture, till the soil, pollinate, fertilize, and most important—plant some seeds and see what grows.

Ask your customer. Maybe they can't identify unmet needs today but they may give you some ideas tomorrow. As Mark Twain said, "The secret of getting ahead is getting started. The secret of getting started is breaking your complex, overwhelming tasks into small, manageable tasks, and then starting on the first one."

Remember that creative energy that started at the very beginning of time and space is within us all.

Stand back and realize that

you are part of this cosmic phenomenon. You are not separate from the universe. It is creatively expanding and you are the current vehicle that allows this creative process to unfold. Within you is infinite creative potential. Limitation is only something stuck in your mind.

To allow it to unfold, simply ask, "What can be done to make conditions around me better? Then listen! Then do it!

"What did you say?"

I said, "**Do it!**"

"Every CEO has to spend an enormous amount of time shuffling papers. The question is, how much of your time can you leave free to think about ideas? To me the pursuit of ideas is the only thing that matters. You can always find capable people to do almost everything else." —Michael Eisner, <u>Fortune,</u> December 4, 1989

Conclusion

"Many organizations have provided suggestion programs for their employees with varying degrees of success. In a Lean organization, the primary objective of the suggestion system is not to solicit the participation of our team members by extracting their good ideas, but to provide a consistent vehicle for teaching individual problem-solving skills. The suggestion system is a *training tool* for individual problem solving. To make this suggestion system work, leaders must commit to helping the team members complete a defined problem-solving process for every idea or suggestion they have. Leaders have to provide access to all the information a particular team member may need to support his or her suggestion. This might include access to engineering, or finance and accounting to quantify the magnitude of the problem and the solution. Leaders commit to responding within 24 hours to every suggestion submitted and to approving the suggestions at the lowest level possible. The supervisor of the person with the suggestion should have the authority to approve the vast majority of suggestions after helping the team member complete the problem-solving process and document the findings on the suggestion form. If we stay focused only on improving the process, we lose sight of the true value of the suggestion system—improving the people.

"So, does focusing on improving people mean we don't measure productivity, cycle times, or costs? Absolutely not. Does focusing on improving people mean we stop doing Kaizen events, accelerated improvement workshops, Lean events, or action workouts (choose your favorite label)? Absolutely not.

197

ALL YOU GOTTA DO IS ASK

Focusing on people doesn't relieve us from the burden of getting results, so we will continue to set goals based on various measures.

"What needs to be different, though, is how we view Kaizen events, suggestion systems, and job design. In a conventional organization, these three activities focus on getting things from the employee (improved productivity, ideas, work) rather than providing something to the employees. If we were to view them instead as tools for improving people, these become learning activities and provide skills and opportunities to employees. Results are still important. But even more important is developing in the workforce the skills needed to sustain improvements. Focus on the people and the results will follow. Focus on the results, and you'll have the same troubles as everyone else—poor follow-up, lack of interest, no ownership of improvements, diminishing productivity. What really needs to be different is attitude."[14]

—David S. Veech, senior staff member in the University of Kentucky Center for Manufacturing.

---

[14] From "A Person-Centered Approach to Sustaining a Lean Environment," by David S. Veech, DAU Press: Publisher. August – November 2004, *Defense Acquisition Review Journal (ARJ)*, *11*(2), p.169. Reprinted from http://www.dau.mil/pubs/arq/2004arq/Veech.pdf with permission.

Appendix

There's Gold in Them Thar Hills!

Discovering the hidden creative potential in all employees

Technicolor in Detroit, Michigan, a Thomson Digital Media Solutions business, has had great success with their employee involvement program through their TIP system (Technicolor Improvement Process) whereby each employee is encouraged to submit at least two improvement ideas per month. Norman Bodek interviewed Pat Goss, Director of Human Resources for Technicolor Michigan, in August, 2003.

Bodek: I would like you to tell me about your TIP program and your HR department's involvement in it.

Goss: Technicolor in Detroit had been heavily involved in implementing Lean manufacturing and **we knew that having all of our employees involved though their creative improvement ideas was really the heart of Lean.** So we at HR were looking to develop better ways for employee involvement and at first, like many other

199

companies, we looked at the traditional suggestion system. We looked at and implemented a very basic system knowing that it was not at all the state of the art but felt that it would at least get us started to involve our employees.

We initially put a lot of work into it, but we soon saw that there was not a lot of participation by employees in relationship to our effort. Unfortunately only a few of the employees were involved. This all changed when we got Chuck Yorke, Manager of Organizational Development, involved in the training process. He offered us a lot of enthusiasm and a strong desire to improve the suggestion process. In fact, once he got turned onto Quick and Easy Kaizen, it was difficult to get him fully back to the traditional other areas of HR.

Chuck attempted to tackle and improve the process and make it more successful for the both employees and the company. Chuck then looked to bring in talented people to help provide us with the experience, knowledge and a fresh way of looking at our TIPs process. Our leader, Mike Karol, read your book, _The Idea Generator—Quick and Easy Kaizen_, passed it on to Chuck, and commissioned him to find ways to bring the technology to Detroit. Chuck contacted you and arranged your first workshop. And it was Chuck's follow through that was a key to make our current program so successful. Imagine last year at this time, we received only 13 ideas from our entire work force and last month we received 1,320 improvement ideas with 687 of them already implemented.

Well, we wanted to create something that would really enrich the lives of the employees and to also give them the recognition for their participation. We wanted the system to become a formalized part of our culture to generate enthusiasm and participation. The rewards generated from the traditional suggestion system were not really effective in getting our employees involved in improvement activities.

Our Lean efforts were getting stuck. There were missing pieces to it. We had visited other facilities and saw how their suggestion process was working very well for them and making them more competitive in this difficult environment. Chuck also was instrumental in moving us through lean and we knew that the employee involvement piece or the TIPs piece was an important part of lean and it was a missing piece. If that's not working the way it should be, then you will not be effective in your lean implementation.

In the last few months, we have begun to see the true impact of Quick and Easy TIPs. It is a very important piece to our lean efforts. Our process changed for us when we began getting away from ideas of "what other people can do for me" to those ideas that people can implement and control themselves. That's when it began to take off. I did pay attention in your class.

By the way, you had a wonderful article with Gary; in some ways it is the best recognition to be recognized by you and others through an article for reference.[15]

Bodek: *What are the benefits to developing human resources through the TIPs system?*

Goss: We want people to be able to control their own destiny. We want people to focus on what they can control, and on what they can be responsible for improving.

---

[15] See Kaikaku The Power and Magic of Lean, PCS Press for the interview with Gary Smuda.

We want them to be willing to try new things. With this Quick and Easy TIPs, people are given new opportunities to control their own work efforts. "It is OK to look at what you are doing and to focus on ideas to improve things around you. It is OK to use a little duck tape and cardboard to try things in new ways and make things a little bit smoother."

It truly has energized people. When you look at the energy developed for them through their own ideas, it has many beneficial effects, not just in the improvement ideas you get on the job but in the person's attitude toward the job, toward the work, and toward coworkers. I would guess it even adds toward the company's reputation in the community, which means we get more people interested in the kinds of things Technicolor is doing.

Truly engaged people have demonstrated to us in many ways that they are a lot more effective when they are allowed to add their own creative ideas into the process. Their personal attitude toward their job completely changes when they see that they do have the freedom to participate. When people participate in improvement activities they are more interested in their work, they have a better day at work, and when you have a good day at work and you are excited about your work, you don't go home and "kick the dog." It might have better effect on employees' family relations as well, and into the rest of their lives. A little employee involvement goes a long way.

Bodek: *I believe that the HR department is burdened with payroll, OSHA, and hiring employees, and might be neglecting one of their prime functions, which is development and training of the workforce. For this reason, many companies are now outsourcing or considering outsourcing the HR function.*

Goss: The HR Department should be, and is in our case, a strategic business partner to provide tools and techniques needed by employees for the overall success of the company to improve productivity, to improve quality, and to reduce costs. You can't do this unless you provide your workers with the knowledge to know what they are doing. The costs of mistakes today are getting higher and higher.

When I first came here to Technicolor we were using a temporary agency to hire new fulltime employees. I couldn't fully appreciate or understand how outside resources would have as much of a stake in improving our culture through good hiring as internal resources. However, since there was lots of expansion and growth going on, we enhanced our HR program to give the kind of support necessary to handle internally all of the HR functions.

Bodek: *What is HR's role in supporting TIPs and the improvement process at Technicolor Corporation?*

Goss: HR's role is to help set objectives and goals and to support the operational staff in obtaining them. We want new employees to quickly fit in and become part of our culture. We want to facilitate the boundary issues and to be sort of the lubricant to keep the wheels moving. We do this through educating employees, and through our reviews and goal setting processes. We want also to have a work recognition program for those people that do have and are willing to offer their good ideas. We do want our employees to have the right tools.

Bodek: *I have been trying to spread Quick and Easy to other companies, but I have found an enormous resistance of managers to try new things. What advice*

203

would you give me to open others to experiment with that which is working so well here?

Goss: I can understand people's reluctance to work with outside consultants, because you have to have some sincere interest among those people that are leading those businesses. But when you want to succeed, you use the best tools necessary no matter where they come from. You have to educate them and take them to places where the system is working. They then have to look at their own cultures and see the benefits of this kind of system and then they have to believe in it.

We hope that our other divisions within Technicolor and Thomson will hear about our success story and want to come and visit us and see things with their own eyes. When they come here and see how it works within our culture then there is a new willingness to go back and attempt to make changes within their own environment.

I have been here now around eight years. We have made much more progress with our improvement activities in the last two years then we made in the first six. Not only through the employee involvement program but our culture has grown and matured and has become more positive and more employee oriented. To answer your question on how to implement it throughout Thomson and Technicolor, the interest has got to be there. When they see our success, they will go back to their sites and say good things are happening in Detroit and we need to find out what it is or we will be left behind. In Technicolor Corporation we are all one family but we do compete for work and when good things happen here and we are having successes it is going to draw the interest from the others. This is a better way to see it grow than to have it mandated.

THERE'S GOLD IN THEM THAR HILLS!

Bodek: You are the HR leader. Without you this would never have happened. Even though Chuck is out there moving it forward, he never would have been able to do this without your support and what is very unique is your willingness to give him the freedom to make this happen. In the long run this will give you an amazing shot in the arm to the lean process.

I thank you for this great interview.

ALL YOU GOTTA DO IS ASK

Biographies

Charles J. Yorke
ChuckYorke@yahoo.com

Chuck Yorke led the process to involve all employees at Technicolor in improvement activities. Over 19,000 improvement ideas were received in the year prior to the publication of this book. Over 10,000 of those ideas were implemented (a 900 percent improvement over two years earlier), resulting in many process and safety improvements as well as significant cost savings. Chuck also implemented an ongoing lean enterprise coaching process for supervisors and hourly associates, promoting the transfer of lean tools and concepts to the plant floor. Prior to his work at Technicolor, Chuck worked as a consultant and trainer. He provided QuestOne Decision Science's Constraints Management training for General Motors Corporation, Ford Motor Company, and other manufacturing organizations. He is also recognized as a certified Kepner-Tregoe program leader for Chrysler Corporation. Chuck has facilitated many workshops in problem solving, communication skills, systems thinking, and other programs. In addition, he has health care experience, with responsibility for operations of facilities with gross revenues over $20 million annually.

Chuck has a Bachelor Degree in Communications from

ALL YOU GOTTA DO IS ASK

Eastern Michigan University, and the following certifications:

- Instructor, situational leadership;
- Coach, The Team Learning Lab (Interactive Learning Laboratories, Inc);
- Instructor, Quick and Easy Kaizen;
- Certified by Development Dimensions International (DDI) as a master trainer for DDI's learning systems;
- Certified by Development Dimensions International (DDI) as a facilitator for over 50 programs including: coaching, trust, leading effective meetings, communication;
- Program leader for Root Cause Analysis (Chrysler);
- Certified by Kepner-Tregoe. Program includes: situation appraisal, problem analysis, decision analysis, and potential problem analysis;
- Instructor, The Power of Focus Management using Outlook (Time/Design, PathWise)..

•

Norman Bodek
Bodek@pcspress.com

Norman Bodek is the president of PCS Inc., a publishing and consulting company in Vancouver, Washington. In 1979, he started Productivity Inc. Press and published hundreds of management books on productivity, quality, and Lean manufacturing. At Productivity, he also developed and published numerous management training programs, training videos, ran major national conferences on productivity and quality, over 200 seminars and workshops a year, 50 study missions to Japan, and published five monthly newsletters. From his 60 trips to Japan, he met and published the works of the Taiichi Ohno, Dr. Shigeo Shingo, the co-creators of the Toyota Production System (Lean manufacturing), and many other masters in manufacturing improvement. He even reprinted and republished the 1926 book of Henry Ford Today and Tomorrow. He introduced to the West important Lean tools and techniques discovered in Japan: JIT, Jidoka, 5S, Value Stream Mapping, SMED, QFD, CEDAC, Hoshin Kanri, Andon, Kanban, and Poka-Yoke. He started the Shingo Prize and is a Shingo Prize winner.

Norman introduced Technicolor Corporation and other companies to Quick and Easy Kaizen. At Technicolor in Detroit, Michigan, prior to his workshops, the company received in the year 2001 only 250 suggestions with 113 implemented, while in the last twelve months (ending October, 2004) the company received 22,000 suggestions with over 12,000 implemented, resulting in over $10,000,000 in cost savings. He is a consultant and frequent speaker on Kaizen, Kaikaku, and Lean manufacturing. In 2004 he was invited to keynote the IIE Lean conference in Los Angeles, California and the Quality System Conference in Detroit, Michigan. In 2005 he will keynote conferences for APICS, ASQ and others. He has written close to 100 published management articles and is the author of two prior books, *The Idea Generator – Quick and Easy Kaizen* and *Kaikaku The Power and Magic of Lean.*

Norman attended the University of Wisconsin and graduated from New York University. Prior to starting Productivity Inc. he was the president of Key Universal Ltd. with offices in Greenwich, CT and Grenada, West Indies.

Bibliography

Abrashoff, D. Michael. *It's Your Ship: Management Techniques from the Best Damn Ship in the Navy.* Warner Books, 2002.

Basadur, Min. *The Power of Innovation: How to Make Innovation a Way of Life and Put Creative Solutions to Work.* Financial Times, Prentice Hall, 1995.

Basadur, Min. *Simplex: A Flight to Creativity.* Creative Education Foundation, 1995.

Bateson, Gregory and Mary Catherine Bateson. *Towards an Epistemology of the Sacred.* Wesleyan University Press, 1979.

Bodek, Norman. *The Idea Generator: Quick and Easy Kaizen.* PCS Press, 2002.

Bodck, Norman. *Kaikaku The Power and Magic of Lean.* PCS Press, 2004.

Bossidy, Larry and Ram Charan. *Execution: The Discipline of Getting Things Done.* Crown Publishing, 2002.

Byham, William C. and Jeff Cox. *Zapp!: The Lightning of Empowerment: How to Improve Productivity, Quality, and Employee Satisfaction.* Ballantine Books, 1998.

Imai, Masaaki. *Gemba Kaizen: A Commonsense, Low-Cost Approach to Management.* McGraw-Hill, 1997.

Imai, Masaaki. *Kaizen: The Key To Japan's*

Competitive Success. McGraw-Hill/Irwin, 1986.

Degraff, Jeff, and Lawrence, Katherine A. *Creativity at Work.* Jossey-Bass, 2002.

Meyer, Pamela. *Quantum Creativity.* McGraw-Hill, 2000.

Ohno, Taiichi. *Workplace Management.* Productivity Press, 1988.

Roberts, Wess, and Ross, Bill. *Leadership Secrets of Star Trek, The Next Generation: Make It So.* Pocket Books, 1996.

Robinson, Alan, and Stern, Sam. *Corporate Creativity: How Innovation and Improvement Actually Happen.* Berrett-Koehler, 1997.

Robinson, Alan, and Schroeder, Dean M. *Ideas are Free: How the Idea Revolution is Liberating People and Transforming Organizations.* Berrett-Koehler, 2004.

Senge, Peter M. *The Fifth Discipline: The Art & Practice of the Learning Organization.* Doubleday, 1990.

Shingo, Shigeo. *The Sayings of Shigeo Shingo: Key Strategies for Plant Improvement.* Productivity Press, 1987.

Taylor, Frederick Winslow. *The Principles of Scientific Management.* Reprint by Dover Publications, 1998.

Tozawa, Bunji. *Kaizen Teian I: Developing Systems for Continuous Improvement Through Employee Suggestions.* Productivity Press, 1992.

Kaizen Teian II: Developing Systems for Continuous Improvement Through Employee Suggestions. Productivity Press, 1992.

Womack, James P. and Jones, Daniel T. *Lean Thinking: Banish Waste and Create Wealth in your Corporation.* Simon & Schuster, 1996.

INDEX

ALL YOU GOTTA DO IS ASK

Examples

Quick and Easy Kaizen	
Before Improvement	**After Improvement**
Parts slip between the gap on the line and the reject basket	Taped a piece of cardboard between the gap and the reject basket to close the gap
The Effect	

No parts fall to the floor and get dirty. Better quality and less time to wipe off the dirt.

Name: Zaid	Date: 1-9-05	Estimated Cost
	Supervisor: Avi	Savings: $ 750

Areas of Improvement: ☐Better Process ☐Customer Service ☐Cost Savings
☐Higher Quality ☐Safety ☐Environmental ☐Other

☐Temporary Fix	☐Long Term Improvement	☐Implemented	☐Not Implemented

Operator estimated that the time saved is 50 hours per year.

Quick and Easy Kaizen	
Before Improvement	**After Improvement**
When sleeving I get paper cuts.	I now wear a glove on my right hand.

The Effect	
No more cuts	

Name: JoAne	Date: 1-7-04	Estimated Cost
	Supervisor: Avi	Savings: $ 22.50

Areas of Improvement: ☐ Better Process ☐ Customer Service ☐ Cost Savings

☐ Higher Quality ☐ Safety ☐ Environmental ☐ Other

☐ Temporary Fix	☐ Long Term Improvement	☐ Implemented	☐ Not Implemented

Instead of stopping work to go to the nurses office, losing time the operator used her own ingenuity.

Quick and Easy Kaizen	
Before Improvement	**After Improvement**
My associates might not be aware of the evacuation procedures in the event of a fire.	Have a practice drill.
The Effect	
My associates will know what to do if there is ever an emergency	

Name: Beverly	Date: 1-12-05	Estimated Cost
	Supervisor: Eli	Savings:

Areas of Improvement: □Better Process □Customer Service □Cost Savings
  □Higher Quality □Safety   □Environmental   □Other

□Temporary Fix	□Long Term Improvement	□Implemented	□Not Implemented

Normally operators just wait until supervision determines when a fire drill will be held. Now we have everyone thinking how to improve safety.

Quick and Easy Kaizen	
Before Improvement	**After Improvement**
Slipcases not being folded correctly making rework for UPC application	Demonstrate the correct way to fold slipcases
The Effect	
Prevent rework on UPC sticker.	

Name: Keith	Date: 1-12-05	Estimated Cost
	Supervisor: Ken	Savings: $ 350

Areas of Improvement: ☐Better Process ☐Customer Service ☐Cost Savings
☐Higher Quality ☐Safety ☐Environmental ☐Other

☐Temporary Fix	☐Long Term Improvement	☐Implemented	☐Not Implemented

Each person can rise to the occasion and take on additional responsibilities if we only ask, respect and listen.

Quick and Easy Kaizen	
Before Improvement	**After Improvement**
Face labels aren't sticking to the cassettes.	Place a towel on trail end of labeler to make label stick to cassettes.
The Effect	
Higher quality and less rework.	

Name: Shawn	**Date**: 1-12-05	**Estimated Cost**
	Supervisor: Mugeeb	**Savings:** $100

Areas of Improvement: ☐Better Process ☐Customer Service ☐Cost Savings ☐Higher Quality ☐Safety ☐Environmental ☐Other

☐Temporary Fix	☐Long Term Improvement	☐Implemented	☐Not Implemented

Sometimes just a simple solution will do.

Quick and Easy Kaizen	
Before Improvement	**After Improvement**
We were waiting for parts. The line was down.	Asked everyone to help me insert new fixtures.
The Effect	
No wasted time.	

Name: Mary Smith	Date: 1-6-05	Estimated Cost
	Supervisor: Alex	Savings: $75.00

Areas of Improvement: ☐Better Process ☐Customer Service ☐Cost Savings ☐Higher Quality ☐Safety ☐Environmental ☐Other

☐Temporary Fix	☐Long Term Improvement	☐Implemented	☐Not Implemented

The operator alerted her team and saved five hours.

Quick and Easy Kaizen	
Before Improvement	**After Improvement**
It was hard to keep the excess shrink-wrap on the spindle because the spindle was very loose.	Wedged a couple of small pieces of cardboard between the spindle and the shrink-wrap roller to keep it tight.

The Effect
Now it is easy to remove the excess shrink-wrap.

Name: Minara	**Date:** 1-9-05	**Estimated Cost**
	Supervisor:	**Savings: $** 15.00

Areas of Improvement: ☐ Better Process ☐ Customer Service ☐ Cost Savings

   ☐ Higher Quality   ☐ Safety   ☐ Environmental   ☐ Other

☐ Temporary Fix	☐ Long Term Improvement	☐ Implemented	☐ Not Implemented

A small but very effective idea showing that the operator was alert to solving problems.

ALL YOU GOTTA DO IS ASK

Quick and Easy Kaizen	
Before Improvement	**After Improvement**
It was hard to get a good grip on the shrink-wrap when trying to pull it off the spindle.	Used a rag to hold on to the shrink-wrap to it off the spindle.

The Effect			
Easier and safer way to take off the shrink-wrap			

| **Name:** Taria | **Date:** 1-9-05 | **Estimated Cost** |
| | **Supervisor:** Anis | **Savings:** $ 15.00 |

Areas of Improvement: □Better Process □Customer Service □Cost Savings □Higher Quality □Safety □Environmental □Other

□Temporary Fix	□Long Term Improvement	□Implemented	□Not Implemented

Solutions are found without having to spend any money at all.

Quick and Easy Kaizen	
Before Improvement	**After Improvement**
Cardboard corners laying on the floor causing people to trip	Placed the cardboard corners up inside the empty recycle bin rack.
The Effect	
Safety, easy to be picked up and cleaner line.	

| Name: Annees | Date: 1-09-05 | Estimated Cost |
| | Supervisor: Anis | Savings: |

Areas of Improvement: ☐Better Process ☐Customer Service ☐Cost Savings

☐Higher Quality ☐Safety ☐Environmental ☐Other

| ☐Temporary Fix | ☐Long Term Improvement | ☐Implemented | ☐Not Implemented |

The old saying that "two heads" are better than one is certainly true. Just imagine when everyone in your company is thinking everyday to improve the process, improve customer service, reduce costs, improve quality, improve safety, improve the environment and other things. You then can understand how it is possible to save thousands of dollars per year per employee from the employees ideas rarely ever tapped into before.

Quick and Easy Kaizen	
Before Improvement	**After Improvement**
I usually open boxes from the top and reach in to take out the parts.	I turned the box upside down and opened it from the other end; easier to take out the parts.

The Effect
Easier and neater way to open the box of parts.

ame: Abdul	ate: 1-10-05	stimated Cost Savings:
	upervisor: Anis	100

Areas of Improvement: ☐ Better Process ☐ Customer Service ☐ Cost Savings

☐ Higher Quality　　☐ Safety　　☐ Environmental　　☐ Other

☐ Temporary Fix	☐ Long Term Improvement	☐ Implemented	☐ Not Implemented

Sometimes the simplest solutions are the best.

Quick and Easy Kaizen	
Before Improvement	**After Improvement**
Parts comes down the line and stacks up and jams up the machine.	Taped an empty plastic spindle across the line to act as a guide to keep parts from stacking up on the line.

The Effect
Better production and less picking up.

Name: Omar	Date: 1-10-05	Estimated Cost Savings:
	Supervisor:	$ 250

Areas of Improvement: ▫Better Process ▫Customer Service ▫Cost Savings
▫Higher Quality   ▫Safety   ▫Environmental   ▫Other

☐Temporary Fix	☐Long Term Improvement	☐Implemented	☐Not Implemented

Many times we can find scrap normally just thrown away to help with our solutions.

Quick and Easy Kaizen	
Before Improvement	After Improvement
I work at the middle of the table but the recycle bin is at the end of the table.	I moved the recycle bin to the middle closer to my work area.

| The Effect |
| Less walking to the recycle bin and easier to empty the recycle bin. |

Name: Ahmed	Date: 1-10-05	Estimated Cost Savings:
	Supervisor: Anis	$ 100
Areas of Improvement: □Better Process □Customer Service □Cost Savings □Higher Quality □Safety □Environmental □Other			
□Temporary Fix	□Long Term Improvement	□Implemented	□Not Implemented

Give the operator an hourly rate to use for cost savings and let them estimate the amount of time they save a year from their ideas. You don't want accountants involved. Just keep it simple and of course, praise, praise and praise.

Quick and Easy Kaizen	
Before Improvement	**After Improvement**
Waiting for the product to come down the line, nothing to do.	I sweep the floor around the machines while waiting.
The Effect	
Using the time wisely and good housekeeping	

Name: Alvin	Date: 1-09-05	Estimated Cost
	Supervisor: Anis	Savings:

Areas of Improvement: ☐Better Process ☐Customer Service ☐Cost Savings
  ☐Higher Quality ☐Safety   ☐Environmental   ☐Other

☐Temporary Fix	☐Long Term Improvement	☐Implemented	☐Not Implemented

Once people are treated with real respect they will rise to the occasion to improve the work environment.

ALL YOU GOTTA DO IS ASK

	Problems	Solutions ?
1	It takes too long for me to put blue dots onto the parts	
2	Line is messy because of damaged parts on tables	
3	I can't change the shrink-wrap roll, boxes blocking	
4	Waiting for labels	
5	QC inspectors not on the line to audit work orders, can't ship	
6	I'm tired of standing on cardboard	
7	Sticker machine not putting label on the right spot	
8	Sometimes the product doesn't fit in the box	
9	Reject basket when full parts fall to floor and get dirty	
10	Labeler was placing sticker on every other box	
11	Skid blocking packing line	
12	Dust is removed by paper towel doesn't always work	
13	Nowhere to set small recycle containers	
14	Some workers can't read the English signs in the workplace	
15	When slow people stand around doing nothing	
16	Bathroom too crowded during breaks	
17	Walls of the break room are very dirty	
18	Only the team leader and supervisor audit the line	
19	One employee inspecting for eight hours	
20	Only lead knows how to run the line	
21	Damaged parts loaded with gook parts onto the line	

ALL YOU GOTTA DO IS ASK

	Problems	Solutions ?
22	Not sure how to write up an idea form	
23	Can't find parking space	
24	Can't find the broom	
25	Can't find the earplugs	
26	Band-aids only available at security desk	
27	WE waited for material for about 90 minutes	
28	People not paying attention at supervisor's meetings	
29	People forget to recharge the batteries at night	
30	No place to put dirty rags	
31	Can't hear the pagers	
32	Need 45 minutes for lunch	
33	Rough edges on yellow railings	
34	No procedures for fire drills	
35	What happened to Kaizen teams?	
36	Don't get feedback on the company's progress	
37	Cable wires dangling from the ceiling and floor for computers	
38	Tacky fingers misplaced	
39	Ink pump not working	
40	No set place for the USDA labels	
41	Not enough quality inspectors	
42	We are not getting enough ideas from people	
43	No incentives given for implemented ideas	
44	No box cutter	
45	Excess plastic drops to the floor	
46	Plastic shrink machine keeps breaking down	

EXAMPLES

	Problems	Solutions ?

ALL YOU GOTTA DO IS ASK

	Problems	Solutions ?

The Idea Generator – Quick and Easy Kaizen

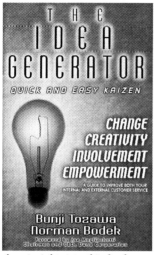

"The Idea Generator" brings continuous improvement down to Earth—and raises up to the heavens the importance of everybody documenting every implemented idea, however small. It's a message too few understand
—**Richard J. Schonberger, President, Schonberger & Associates, Inc.** and author of *Japanese Manufacturing Techniques: Nine Hidden Lessons in Simplicity*

Lean systems will degrade without ongoing improvement from every employee through a myriad of simple, quick changes. What brings lean structures to life is people—people engaged in continuous improvement. Tozawa and Bodek provide deep insights into this fundamental ingredient of high performance companies
—**Dr. Jeffrey K. Liker, University of Michigan and author of Becoming Lean**

Kaikaku — The Power and Magic of Lean

"Norman has always had fascinating stories about his travels around the world, the people he's met, and the things he's learned. Some of the most fascinating of these stories concern his friendship with Mr. Shigeo Shingo. As you read through Kaikaku, it is my sincere hope that you will appreciate the years of learning that went into its creation, as well as the great many things that Norman has brought to this country, and the world."
—**Bill Kluck, President, NWLEAN.net**

One of the things you mentioned in your email to me was that you recommend I read Kaikaku. I purchased it a couple months ago and it just now made it to my reading rotation. If I had only known what a great book it is I would have devoured it the moment it arrived!
—**James A. Feltman, Sales Manager, Vorne Industries, Inc.**

ALL YOU GOTTA DO IS ASK